CW00391237

Concise
Dictionary
of
Biology

Concise
Dictionary
of
Biology

KEITH WHITTLES BSc, PhD, FGS
ALICE GOLDIE BSc

**TIGER BOOKS INTERNATIONAL
LONDON**

© Geddes & Grosset Ltd 1993

This edition published in 1993 by
Tiger Books International PLC, London
Reprinted 1994.

ISBN 1-85501-366-5

Printed and bound in Slovenia

CONTENTS

A

abdomen area of the animal body that lies posterior to the THORAX. In mammals it is divided from the thorax by a DIAPHRAGM and in the vertebrates as a whole contains VISCERA (e.g. LIVER, KIDNEYS, INTESTINES).

ABO blood group *see* **blood grouping**.

absolute temperature a temperature measured on the KELVIN SCALE with respect to ABSOLUTE ZERO.

absolute zero the temperature at which the particles of matter have no thermodynamic energy, and which has been given the theoretically lowest value of -273.15°C (-459.67°F).

abyssal zone refers to the environment of the deep ocean floor extending up to approximately 2000 metres (6500 feet). There is virtually no light penetration and organisms are suitably adapted for life in a dark, cold environment, under great pressure and often with little food.

acacia gum *see* **gum arabic**.

acetic acid more correctly, ethanoic acid (CH_3COOH) is obtained by the oxidation of ethanol or synthesized from acetylene. It is the acid in vinegar (an aqueous solution of 3 to 6 per cent).

acetone *or* **propanone** (dimethyl ketone CH_3COCH_3) an important solvent used much in organic synthesis, e.g. in making plastics.

acetylcholine one of the important substances in the nervous system, transmitting impulses from one nerve to another (*see* SYNAPSE).

acetyl salicylic acid *see* **aspirin**.

acid any substance that releases hydrogen ions during a chemical reaction, thus lowering the pH of the solution. An acidic solution has a pH of less than 7 and will react with a BASE to form a salt and water. An acid can be thought of as a proton (H^+) donor (Brönsted-Lowry acid) or as an electron pair acceptor (Lewis acid).

acid rain rain with a high concentration of pollutants, such as dissolved sulphur and nitrogen oxides, that have harmful effects on plant and animal life. The pollutants are principally byproducts of industrial activities involving the burning of coal or oil (*see* FOSSIL FUELS).

acquired immune deficiency syndrome *see* **AIDS**.

acquired immunity *see* **adaptive immune system**, **immune system**.

ACTH (*abbreviation for* adrenocorticotrophic hormone, corticotrophin) a protein hormone secreted by the anterior PITUITARY GLAND. It is released in response to stress and acts upon the adrenal glands (*see* ENDOCRINE SYSTEM), thus controlling the release of the CORTICOSTEROID hormones. It is used medically as an anti-inflammatory agent to relieve the symptoms of such conditions as asthma and rheumatic diseases.

actin a protein with globular and fibrous (polymerized) forms, found in muscle with MYOSIN, and now known to be associated with sites of cellular movement.

actinomycetes members of an order of bacteria (Actinomycotales) which tend to occupy ANAEROBIC conditions and be static. All species are like FUNGUS in that their cells are ordered in a HYPHA-like arrangement of delicate filaments. An important genus, *Actinomyces*, contains some species that cause diseases in both man and animals. The *Streptomyces* are medically important as a source of antibiotics such as streptomycin.

Actinozoa *or* **Anthozoa** a class of animals belonging to the phylum Coelenterata (now phylum CNIDARIA) including sea pens, sea pansies, sea feathers, sea fans, sea anemones and corals. These animals are all marine and do not have the free-swimming, medusoid stage (*see* MEDUSA) in the life cycle. The body is composed of a POLYP which may be single or colonial. The corals produce a calcareous EXOSKELETON while others have an internal skeleton.

action potential a change in the electrical potential that occurs across the cell membrane (between the inside and outside), of a nerve AXON as an IMPULSE travels along. It is due to the nerve cell receiving a stimulus and is a temporary and localized occurrence which travels down the nerve axon in a wave-like manner. This voltage pulse is caused by sodium ions entering the axon causing a change in potential across the cell membrane from -65mV (millivolts) to + 45. Once the stimulus has passed, the cell membrane is restored to its resting electrical potential. If a stimulus is continually received, repeated pulses will travel along the nerve at a rate of several hundred per second. Similar effects to those described occur in a muscle fibre which receives nervous stimulation and in the latter case leads to a continuous response in the muscle (tetanus).

activation energy the input of energy required to

initiate a chemical reaction. In some reactions, the KINETIC ENERGY of the reactant molecules is sufficient for the reaction to proceed, but in others the activation energy is too high for the reaction to occur (*see* CATALYST).

active site the place on the surface of an ENZYME molecule which, because of its particular three-dimensional arrangement and distribution of charge, gives it the ability to bind to its substrate molecule. The properties of an active site are governed by the POLYPEPTIDE chains of the enzyme, and the AMINO ACIDS of which these are composed. A temporary enzyme-substrate complex is formed by the binding of the active site to the substrate molecule. More than one such reaction may be catalysed (*see* CATA-LYST) as an enzyme may have a number of active sites. Inhibitor molecules of the normal substrate reaction may compete and bind to the active site so its availability for normal reaction is reduced. Enzyme activity can be subject to INHIBITION in this way.

active transport the transport of substances across a cell membrane against an electrochemical potential gradient or concentration gradient. The process requires an input of energy which is usually ATP (adenosine triphosphate) but may be derived from other sources such as a gradient of protons across the cell membrane. Inorganic ions and organic molecules are carried both into and out of cells by means of "pumps" composed of protein molecules that are located in the membrane itself. The transported substance is then released on the other side of the cell membrane. It is probable that all cells carry out active transport and it is mainly used to keep the ion balance in cells correct. The most obvious example is the maintenance of the sodium and potassium ion concentration gradients in nerve and muscle cells.

adaptation a feature of an organism that has evolved under natural selection as it enables the organism to function efficiently in a specific aspect of its particular environmental NICHE. The adaptation can be either psychological, where exposure to certain environmental conditions will only cause a change in the organism's behaviour, or it can be genetic, as when the organism possesses GENES that produce characteristics that prove to be beneficial for survival in its environment. Examples of genetic adaptation are the genetic changes that occur in bacterial populations, conferring antibiotic resistance, as in the strains of bacteria resistant to penicillin. Such antibiotic resistance is a major problem in hospitals when trying to prevent any risk of post-operative infections.

adaptive immune system one functional division of the IMMUNE SYSTEM that produces a specific response to any PATHOGEN. There are two mechanisms of acquired or adaptive immunity:

(1) Humoral—the production of soluble factors, called ANTIBODIES, by activated B-CELLS.

(2) Cell-mediated immunity—the range of T-CELLS involved in the specific recognition of an antigen, which must be present on the surface of another cell, i.e. an antigen-presenting cell (APC). The immunity produced, **adaptive immunity**, is also known as **acquired immunity** because the adaptive immune system is capable of remembering any infectious agent that has induced the proliferation of B-cells (it has acquired a memory for that particular agent, *see* B-CELLS).

adaptive radiation the evolutionary separation of species into numerous descendent species in order to exploit the various habitats that exist throughout the world. This evolutionary divergence of species probably explains the bewildering array of, say, amphibians, which has arisen as a result of adaptive radiation after the first amphibians moved on to land.

adenine ($C_5H_5N_5$) a nitrogenous base component of the NUCLEIC ACIDS, DNA and RNA, which has a PURINE structure. In DNA, adenine will always base pair with THYMINE, but in RNA, during TRANSCRIPTION, adenine will base pair with URACIL. Adenine is also present in other important molecules, such as ATP.

adenosine a nucleoside composed of one adenine molecule attached to D-RIBOSE sugar molecule. The phosphorylated derivatives of adenosine (AMP, ADP and ATP) are of critical importance to the function of living organisms as carriers of chemical energy.

adenosine triphosphate *see* **ATP**.

adipose tissue a connective tissue in the body made up of cells containing fat. There are two types:

(1) Brown adipose tissue (brown fat) is composed of cells with granular CYTOPLASM and is found mainly between the shoulder blades and around the neck of neonatal mammals and mammals that hibernate. The functional role of brown fat is thought to be in the release of heat but it is not common in most adult mammals. It has a rich blood and nerve supply.

(2) White adipose tissue has a poor nerve supply and has a wide distribution in animal bodies, especially in the SUBCUTANEOUS tissue below the skin and around

major organs such as the heart and kidneys. Each fat cell is large, containing a single fat droplet and only a thin layer of cytoplasm. The fat deposits provide an energy reserve for the body and can be utilized by the activation of certain metabolic activities. There is also an important insulating and protective function.

ADP *see* **ATP**

adrenal gland *see* **endocrine system**.

adrenaline (epinephrine, American) a HORMONE derived from the AMINO ACID tyrosine which is secreted by the inner part of the adrenal glands (the MEDULLA). The release of adrenaline prepares the body for "fright, flight or fight" by increasing the depth and rate of respiration, raising the heart beat rate and improving muscle performance. It also has an inhibitive effect on the process of digestion and excretion. It is secreted to a lesser extent by the SYMPATHETIC NERVOUS SYSTEM functioning as a NEUROTRANSMITTER. It can be used medically in the treatment of bronchial asthma, and also during surgery may be administered to constrict blood vessels, thereby reducing bleeding.

adsorption the taking up, or concentration, of one substance at the surface of another, e.g. a dissolved substance on the surface of a solid.

aerobic respiration a set of ENZYME-catalysed reactions requiring oxygen to release energy during the degradation of glucose. The first stage of aerobic respiration, GLYCOLYSIS, occurs in the CYTOPLASM of both plant and animal cells. However, the remaining two stages occur in the MITOCHONDRIA of EUCARYOTIC cells and the membrane of PROCARYOTIC cells. Thirty-eight molecules of ATP are generated per molecule of glucose undergoing complete aerobic respiration.

aerosol a fine mist or fog in which the medium of dispersion is a gas. Also, a pressurized container with spray mechanism used extensively for deodorants, insecticides, etc.

afferent carrying or leading towards. Applied to types of blood vessels or nerve fibres, e.g. sensory nerves. Opposite of EFFERENT.

aflatoxins four related compounds that are produced as a result of secondary metabolic processes by the moulds *Aspergillus flavus* and *A. parasiticus*. They can be highly toxic to cattle and to man if stored foodstuffs on which the mould has grown are consumed. The mould can grow on peanuts and cereals such as rice, and also cotton seed. Aflatoxins prevent replication of DNA by binding to it and cause severe liver damage and cancers. They are a probable cause of liver cancer in Africa.

agar derived from the cell walls of certain species of red algae (e.g. some seaweeds), is widely used as a gelling agent. It is a component of many foodstuffs, cosmetic preparations and medicines. An important biological use is nutrient agar made from blood or beef extract gelled with agar. In microbiology this is used for the cultivation of bacteria, some algae and fungi.

Agnatha a class (or sometimes a super-class) of vertebrates comprising freshwater and marine eel-like animals that have no jaws or pelvic fins. The mouthparts are adapted for sucking and have horny teeth and the skeleton is cartilaginous. The only living forms are the lampreys and hagfishes which belong to the order Cyclostomata. These lead a scavenging or parasitic (*see* PARASITE) existence, and are the remnants of a very ancient fossil group. The earliest known fossil vertebrates are agnathans which had a bony plate-like armour and occur in rocks dated 440–345 million years old, during the SILURIAN and DEVONIAN geological periods.

AIDS (*acronym for* Acquired Immune Deficiency Syndrome) a serious disease thought to be caused by a human retrovirus called HIV—human immune deficiency virus—which kills the adult T-CELLS of the IMMUNE SYSTEM. This destruction of a critical aspect of the body's immune system leaves the patient open to both minor and major infections, as well as the possibility of developing cancer. Not all HIV-infected individuals develop AIDS, but it can be passed on by the following methods: the receiving of infected blood during transfusions; a mother passing it on to a foetus; the sharing of needles in drug abuse; and the passing of body fluids during sexual contact. There is no evidence that HIV can be transmitted by everyday activities and social contact, such as swimming, sharing cutlery, using a public toilet, etc.

akaryote a cell without a nucleus or one where the protoplasm of the nucleus does not form a discrete nucleus.

albinism describes the condition in which pigment, particularly MELANIN, fails to develop in the skin, iris of the eye and hair, resulting in the animal or person being an *albino*. An albino has very light skin, white hair and pink eyes (due to reflection from blood capillaries behind the retina). The condition is hereditary and the ALLELE responsible (*see also* GENE) is recessive to the one which produces normal pigmentation and hence is homozygous (*see* HOMOZYGOTE) in the affected individual.

albumin describes a group of small PROTEINS that are soluble in water and globular. If heated they form insoluble coagulates and are found in the egg white of some reptiles and birds, where the term albumen describes the protein component. Albumins are also found in plants, milk and blood plasma—serum albumin—where they form about 55 per cent of the total protein. Here they help to regulate osmotic pressure (*see* OSMOSIS) thus also plasma volume. They also have a major role in binding and transporting free FATTY ACIDS.

alcohol an organic compound similar in structure to an ALKANE but with the functional group -OH (hydroxyl) instead of a hydrogen atom. Alcoholic beverages contain ETHANOL, the alcohol obtained during the fermentation of sugars or starches.

aldehyde (*also known as* **alkanal**) any of a class of organic compound with the CO RADICAL attached to both a hydrogen atom and a hydrocarbon (alkyl) radical, giving the type formula R.CO.H.

aldosterone an important CORTICOSTEROID hormone produced by the zona glomerulosa of the cortex of the ADRENAL GLANDS. It controls sodium excretion by the kidneys by promoting retention of sodium ions and water. Hence it regulates the salt/water balance in the body fluids.

alga (*plural* **algae**) the common name for a simple water plant, which is without root, stem or leaves but which contains CHLOROPHYLL. Algae range in form from single cells to plants many metres in length. The blue-green algae, CYANOBACTERIA, are widely distributed in many environments.

alimentary canal (alimentary tract, digestive tract, gut) a tube that widens and narrows along its length in which all the food ingested by an animal is processed. In some simple animals such as flatworms and jellyfish, there is only one opening which is used both for the intake of food and for the elimination of waste. However, in most animals the tubular alimentary canal opens at the mouth, where food is ingested, and also at the ANUS where any indigestible material left is passed out of the body. In between, it is divided into different areas which variously specialize in breaking down the food (digestion) and absorbing the products into the body (absorption). There are many glands involved and usually muscular activity moves the food along. Also, the alimentary canal is well supplied with blood vessels (*see also* STOMACH, DUODENUM etc).

aliquot a small sample of material analysed as being representative of the whole.

alkali a soluble base that will give its solution a pH value greater than 7. The hydroxides of the metallic elements sodium (Na) and potassium (K) are strong alkalis, as is ammonia solution (NH_4OH).

alkaloids basic organic substances found in plants and with a ring structure and the general properties of amines. In nature, they act as deterrents for the plant against HERBIVORES. Many alkaloids are used in medicine, e.g. cocaine, codeine, MORPHINE, QUININE.

alkanal *see* **aldehyde**.

alkane an open-chain HYDROCARBON with single bonds between each carbon atom. The first member of this HOMOLOGOUS SERIES is METHANE, CH_4, and the subsequent members may be considered to be derived from methane by the simple addition of the unit, -CH_2. All the members conform to the general formula of C_nH_{2n+2}, thus the chemical formula for the second member of the series, ETHANE, is C_2H_6. Alkanes are SATURATED COMPOUNDS as all the valence electrons of the carbon atoms are engaged within the single, COVALENT BONDS. As the alkanes are saturated compounds, they are quite stable but will undergo a slow substitution reaction with a halogen. For example:

CH_4 (g) + Cl_2 (g) ——> CH_3Cl (g) + HCl (g)
methane chlorine   chloromethane hydrogen chloride

alkene an open-chain HYDROCARBON containing a carbon-carbon double bond. The first member of this HOMOLOGOUS SERIES is ETHENE, C_2H_4, and all other members conform to the general formula C_nH_{2n}. Alkenes are UNSATURATED hydrocarbons, which can easily undergo an addition reaction as each carbon atom has one available ELECTRON that is not engaged in the formation of the double bond.

alkyne an open-chain HYDROCARBON containing a carbon-carbon triple bond somewhere in its molecular structure. The first member of the alkynes is ETHYNE (also called acetylene), C_2H_2, and the general formula for the other members is C_nH_{2n-2}. They are UNSATURATED compounds that will readily undergo addition reactions across their triple bond, due to the availability of the four electrons engaged in the formation of two of the carbon-carbon bonds of the triple bond. The remaining carbon-carbon bond of the C=C bond and any carbon-hydrogen bond are a much stronger type of bond, which makes them more stable and less reactive.

allantois one of the three extraembryonic membranes that protect the embryo. In man the others are the CHORION and AMNION. Specifically in birds, reptiles and mammals the allantois is the means whereby the embryo receives oxygen and in all but the mammals (for which it also helps to supply nutrients) it stores waste products in the egg.

allele one of several particular forms of a GENE at a given place (locus) on the CHROMOSOME. Alleles, responsible for certain characteristics of the PHENO-TYPE, are usually present on different chromosomes. It is the dominance of one allele over another (or others) that will determine the phenotype of the individual.

allergy (*plural* **allergies**) an overreactive response of the IMMUNE SYSTEM of the body to foreign ANTIGENS. Allergies result from the hyperproduction of one class of ANTIBODY, IgE, which activates the release of certain products, including HISTAMINE, by the MAST CELLS. This causes the characteristic symptoms of an allergy, i.e. inflammation, itching, etc. An induced overreaction by the immune system can be produced by an environmental substance like pollen, certain foods, or even toxins injected by insects such as wasps. Such reactions can be counter-attacked using antihistamine drugs.

alpha-napthol test (Molisch's test) describes a test devised by H. Molisch (1856–1937), an Austrian chemist, to discover whether carbohydrates are present in a solution. The test solution is mixed with a little alcoholic alpha-napthol and then concentrated sulphuric acid is slowly added down the side of the test tube. If carbohydrates are present a violet coloured ring is formed where the two liquids meet.

altruism refers to a type of animal behaviour in

which an individual puts its own survival at risk or decreases, or even sacrifices, its chances of breeding in favour of another individual. Altruism apparently conflicts with Darwinian theory, but in many cases seems to have evolved because the individuals concerned are closely related. In animals, it is considered to be an evolved behaviour and not a conscious act of goodwill on the part of the individual that exhibits it, and confers advantages in terms of overall survival.

alveolus (*plural* **alveoli**) descriptive term for a small sac, cavity or depression depending upon the biological situation and/or the species. Most familiar are those collected together to form the alveolar sacs which are found in the lungs of vertebrates at the ends of the bronchioles (*see* BRONCHUS). Each alveolus has a rich supply of blood CAPILLARIES and is lined with a thin moist membrane where respiratory gases (oxygen and carbon dioxide) are exchanged. In total a vast surface area is provided by all the alveolar sacs so that respiration functions effectively.

amides organic compounds in which the hydroxyl (-OH) from the carboxyl group (-COOH) in acids has been replaced by the amide group $-NH_2$. In effect they are ammonia derivatives with the general formula $RCONH_2$.

amines compounds formed from ammonia, NH_3, by the replacement of one or more hydrogen atoms with organic RADICALS. There are three classes: primary, NH_2R; secondary, NHR_2, and tertiary, NR_3.

amino acid any of the 20 standard organic compounds that serve as the building blocks of all PROTEINS. All have the same basic structure, containing an acidic carboxyl group (-COOH) and an amino group ($-NH_2$), both bonded to the same central carbon atom referred to as the α-carbon. Their different chemical and physical properties result from one variable group, the side chain or R-group, which is also attached to the α-carbon.

amino group an essential part ($-NH_2$) of the AMINO ACIDS.

ammonia a COVALENT compound (NH_3) that exists as a colourless gas. It will react with water, giving the alkaline solution known as ammonium hydroxide (NH_4OH). Ammonia will ionize in water to form the ammonium ion (NH_4^+) and the hydroxide ion (OH^-)

$$NH_3 \text{ (g)} + H_2O \text{ (l)} \longrightarrow NH_4^+\text{(aq)} + OH^-\text{(aq)}.$$

ammonite a fossil subclass of cephalopods which were abundant from the DEVONIAN to the Upper CRETACEOUS. All species are now extinct. The shell, which was commonly coiled, contained chambers and often exhibited outer marks and patterns.

amniocentesis medical procedure carried out to sample the amniotic fluid in the womb of a pregnant woman. Analysis of the fluid can determine the condition of the foetus by detecting such conditions as spina bifida and DOWN'S SYNDROME. It is collected by means of a hollow needle which is inserted into the abdomen through the wall of the womb (uterus).

amnion a fluid-filled sac enclosed by a membrane in which a reptile, bird or mammalian embryo develops. The development of the amnion was the evolutionary step which freed animals to breed away from water as it provided the necessary fluid-filled environment for the embryo.

amniote reptiles, birds or mammals which all develop extraembryonic membranes (CHORION, ALLANTOIS) that enclose the AMNION in which the embryo grows (*see* ANAMNIOTE).

amorphous having no shape or form—non-crystalline.

Amphibia a class of vertebrates comprising salamanders, newts, frogs and toads. These were the first vertebrates to colonize the land and evolved in the DEVONIAN (about 370 million years ago). Modern forms tend to be highly specialized and not typical of their fossil ancestors, and many of their characteristics are adaptations for life on land. Most have to return to water to breed and lay ANAMNIOTE eggs in clusters, which are fertilized externally. However, in some, internal FERTILIZATION occurs, and there are also types that are VIVIPAROUS. The young, or larvae, are aquatic, and have GILLS through which they breathe. In order to become adults they undergo METAMORPHOSIS. Adults respire through nostrils that are linked by a passage to the roof of the mouth, and also through their skin, which is kept moist.

amphoterism a property of the oxides or hydroxides of certain metallic elements, which allows them to function both as acids and alkalis. Although insoluble in water, amphoteric compounds will dissolve in acidic solutions ($pH < 7$) or basic solutions ($pH > 7$).

anabolic steroid anabolic refers to ENZYME reactions in the body which cause more complex substances to be formed out of simpler ones, and steroids are a group of chemically similar LIPIDS that have a nucleus of four carbon rings. ANDROGENS, the male sex hormones, are naturally occurring anabolic steroids. Anabolic steroids increase the growth of tissues, especially the muscles, and synthetic forms are used medically to promote weight gain. Misuse by athletes is banned as it can cause long-term damage to the liver.

anabolism the biosynthetic building-up processes of METABOLISM. For example, the synthesis of fatty acids and phospholipids in the GOLGI APPARATUS.

anaemia *see* **erythrocyte**.

anaerobic respiration a set of ENZYME-catalysed reactions releasing energy from the SUBSTRATE in the absence of oxygen. The first stage of anaerobic respiration is GLYCOLYSIS—as in AEROBIC RESPIRATION—but there is only one other stage, FERMENTATION, which has two alternative pathways as follows:

(1) The production of lactic acid, as in muscle cells during prolonged contraction.

(2) The production of the alcohol, ethanol, by micro-organisms such as yeast or some plants.

In all anaerobic respiration, only two ATP molecules are generated (during glycolysis) per glucose molecule.

anamniote more primitive vertebrates—agnathans (*see* AGNATHA), fish and amphibians—which do not produce an AMNION for the development of the embryo.

anaphylaxis an abnormal reaction of the immune system, which may be mild or extremely severe. It occurs when an individual is re-exposed to a certain ANTIGEN that he or she has encountered before (perhaps following an injection of a drug or a bee sting), and the immune response is a release of HISTAMINE and other similar agents. Mild symptoms might be a rash, but severe ones are pallor, breathing difficulties and a rapid drop in blood pressure, possibly leading to unconsciousness, heart failure and death.

androgen one of several male sex hormones of which the most important is TESTOSTERONE. Androgens are mainly secreted by the testes (*see* TESTIS) and are responsible for the growth of these and for the development of the male secondary sexual characteristics—growth of pubic, facial and chest hair and deepening of the voice. There is also a small production of androgens by the ADRENAL GLANDS and female ovaries (*see* OVARY).

Angiospermae (*alternatively* **Anthophyta**) the class of flowering plants. They are the most complex and highly developed plants, in terms of their structure, and are able to exploit a great many different habitats, having evolved various specializations. The female reproductive cell (gamete) is formed within a structure called an OVULE, itself protected by a closed sheath, the CARPEL. After fertilization, the ovule develops into a seed, and seeds are also borne within fruits. There are two classes, the MONOCOTYLEDONS and the dicotyledons. They characteristically show double fertilization.

angström the unit of measurement (10^{-10}m), now superseded by the nanometre ($10\text{Å} = 1\text{nm}$), which was used for electromagnetic radiation.

anion a negatively charged ion formed by an atom or group of atoms gaining one or more electrons.

Annelida an invertebrate phylum containing three classes—POLYCHAETA (e.g. ragworms), OLIGOCHAETA (e.g. earthworms) and Hirudinea (e.g. leeches). The CUTICLE (outer layer) is usually cartilaginous bearing stiff bristles (chaetae) made of CHITIN. Annelids are cylindrical in shape with soft bodies divided into (metameric) segments, and marine, fresh water and terrestrial forms exist. A fluid-filled cavity, the COELOM, provides a hydrostatic skeleton, against which circular and longitudinal muscles in the body wall contract to effect movement. The central nervous system consists of two pairs of ganglia (the brain) lying DORSAL and VENTRAL to the oesophagus and connected by links known as commissures. The brain is connected to a pair of nerve cords that run longitudinally beneath the gut along which lie ganglia in each segment.

annual a plant that germinates from seed, grows flowers and produces seeds within one year, following which it dies. Many familiar examples exist, e.g. cornflowers. (*See also* PERENNIAL, EPHEMERAL).

annulus (*plural* **annuli**) in animals, any ring-shaped structure such as the segments of annelids. In botany it describes a ring of tissue that is left on the stalk of a toadstool or mushroom formed from a ruptured membrane that once covered the cap. It also describes a particular area of thickened cells in fern SPORANGIA and moss capsules, which operates during the release of spores.

anode the positively charged electrode of an electrochemical cell to which the ANIONS of the solution move to give up their extra electrons, i.e. the electrode at which OXIDATION occurs.

ANS *abbreviation for* AUTONOMIC NERVOUS SYSTEM.

antagonistic action an action in which two or more systems or processes act in opposition to each other such that the activity of one reduces or cancels out that of the other. Two muscles may operate in this way where contraction of one must involve relaxation of the other, e.g. in the movement of a limb. Also HORMONES and drugs can act antagonistically such that the release of one inhibits the effects of another. Drug treatments for cancer often operate in this way by inhibiting ENZYME activities of cancer cells.

antenna (*plural* **antennae**) one of a pair of usually long, highly mobile and segmented projections that occur on the head of most ARTHROPODS. Commonly they are very important sense organs, connected with smell and touch, but they may be greatly altered and specialized in some insects. In crustaceans (*see* CRUSTACEA), they are modified for attachment and swimming and another pair of appendages is present to perform the sensory function.

anther part of the male reproductive organ of a flowering plant (*see* STAMEN) in which pollen is contained within pollen sacs.

Anthophyta *see* **Angiospermae**.

Anthozoa *see* **Actinozoa**.

antibiotic a chemical produced by micro-organisms, such as BACTERIA and moulds, that can kill bacteria or prevent their growth. The first antibiotic to be discovered was penicillin, and there are now many more, including erythromycin, streptomycin and terramycin.

antibody (*plural* **antibodies**) a protein, circulating in the blood, which is produced by B-CELLS and will bind to the surface of an ANTIGEN. The production of antibodies is a specific immune response of the ADAPTIVE IMMUNE SYSTEM. Antibodies consist of protein chains that form IMMUNOGLOBIN, i.e. Ig, and are very useful for identifying specific types of protein unique to a particular plant or animal or to a VIRUS that may be circulating throughout the body of an individual. Although millions of different antibodies are produced in order to cope with any PATHOGEN that may arise, there are only five major classes, which have the following functions:

Antibody Function

IgG	The most abundant, it combats micro-organisms and toxins. As it can cross the placenta, it is the first Ig found in newly born infants.
IgA	The major Ig in mucosal secretion, it defends external surfaces including the gut wall to help cope with antigens found within the gut.
IgM	The first Ig to be produced during infection, it is very effective against bacterial infections.
IgD	This is present on surfaces of B-cells, but no specific function is known for it.
IgE	This protects external surfaces and triggers the release of histamine from other cells of the immune system.

anticoagulant any substance which prevents blood from clotting and occurs naturally in the body as, for example, heparin. Anticoagulants in the body help to regulate the blood clotting process that is involved in the healing of wounds. Blood-sucking animals—vampire bats, leeches, insects—often produce anticoagulants which are injected into the wound with the saliva and enable the blood meal to be obtained more easily. Anticoagulants are important medically in the treatment of embolism and thrombosis, and to prevent collected blood from clotting.

antigen any substance that triggers an immune response because of the body's IMMUNE SYSTEM recog-

antihistamine

nizing it as foreign. Common antigens are proteins present on the surface of bacteria and viruses. Unsuccessful transplant operations are usually a result of the patient's immune response recognizing the surface cells of the organ from the donor as non-self. The organ is said to be rejected when the patient's immune system becomes activated and tries to destroy the donated organ.

antihistamine a substance or drug which counters the effects of HISTAMINE release in the body. Hence antihistamines are medically important in the treatment of allergic reactions such as hay fever.

antiseptic a substance used to inhibit the growth of, or kill, micro-organisms, especially BACTERIA, that are the cause of disease. Antiseptics are widely used on living tissue, and especially on the skin, in the treatment of minor wounds and a common one is hydrogen peroxide.

anus the opening of the ALIMENTARY CANAL, other than the mouth, through which the waste products of digestion are expelled, and generally occupying a posterior position in the body.

aorta the largest ARTERY in the body, which acts as the blood outflow of the left VENTRICLE of the heart. The aorta has an approximate diameter of three centimetres, with thick muscular walls to carry blood under pressure. The aorta divides into several branches, which supply blood to the arms and the head. It then continues down around the spine to the level of the lower abdomen, where it divides into two major branches to supply the legs.

apatite a calcium phosphate mineral, also containing fluorine, chlorine, and HYDROXYL ions. It is the primary constituent of fossil bones and vertebrate

28

teeth. It is used industrially in the manufacture of fertilizers.

APC *abbreviation for* antigen-presenting cell. *See* ADAPTIVE IMMUNE SYSTEM.

apoenzyme an ENZYME that is inactive until it combines with a particular COFACTOR to function.

appendage a protrusion from an animal body, often paired, that performs a particular function. Examples are ANTENNAE and paired walking legs in ARTHROPODS, the FINS of fish and the limbs of vertebrates.

appendix *or* **vermiform appendix** a small outgrowth of the part of the alimentary canal known as the caecum. It contains lymphatic tissue but does not have a role in the digestive process. Inflammation of the appendix causes the condition known as appendicitis.

aqueous humour a watery fluid present in the vertebrate EYE between the CORNEA and the lens, that supplies nutrients to these structures and has a composition similar to CEREBROSPINAL FLUID. Its pressure helps to maintain the shape of the eye and it is produced continuously, being completely renewed every four hours.

Arachnida a class of mostly terrestrial arthropods. The head and thorax are not divided, forming a *prosoma*, usually made up of 8 segments. The rest of the body forms the *opithosoma*, generally composed of 13 segments. On the prosoma, the first pair of APPENDAGES are for piercing and grasping (chelicerae) and the second pair (the pedipalps) can be specialized to perform a variety of different functions (as sense organs, for copulation or for noise production). These are followed by four pairs of walking legs. On the opithosoma specialized appendages may also occur such as those used for silk-spinning in spiders. Arachnids are mainly carnivorous and many use poison to immobilize their prey, via the chelicerae in spiders and the sting in scorpions. There are several orders including mites and ticks, scorpions, pseudo-scorpions, spiders, harvestmen and king crabs.

archaebacteria an ancient line of bacteria with distinctive and unusual biochemical characteristics that make them very different from other types of bacteria. Many grow best at temperatures around 100°C and are favoured by an acidic, hot environment, and they may be AEROBIC or ANAEROBIC.

aromatic hydrocarbon any compound that has a molecular structure based on that of BENZENE. Aromatic hydrocarbons are UNSATURATED, closed-chain compounds that will undergo substitution reactions as well as addition reactions, depending on which functional groups are present within their structure and the reactivity of the other reagents.

arteriole a small blood vessel with a muscular wall that transports blood from the ARTERIES to the CAPILLARIES.

arteriosclerosis the thickening, hardening and loss of elasticity of the ARTERIES. This can be a pathological condition of advancing age, or it can be associated with fatty deposits, particularly CHOLESTEROL, that block the arteries, causing their diameter to decrease. As a result, the heart must strain to increase its muscular activity to generate enough pressure to pump the blood through the arteries.

artery (*plural* **arteries**) a thick-walled vessel that carries blood under pressure resulting from the pumping mechanism of the heart. The PULMONARY ARTERY carries deoxygenated blood from the heart to the lungs, but all other arteries carry oxygenated blood from the heart to body tissues.

Arthropoda the largest phylum of invertebrate animals containing over one million species which occupy numerous different habitats and are often highly specialized. There is generally a hard outer layer (EXOSKELETON) comprising a CUTICLE made of CHITIN. The animal grows by shedding this periodically (*see* ECDYSIS). The arthropod body is segmented and three distinct regions are usually produced; the head, thorax and abdomen. The segments often have APPENDAGES that perform a variety of functions. The main classes include the Onychophora, MYRIAPODA (millipedes and centipedes), INSECTA (insects) CRUSTACEA (lobsters, crabs, shrimps) and ARACHNIDA (spiders, mites, scorpions).

Artiodactyla made up of one order of mammals, the Ungulates, which are even-toed having the third and fourth toes equally developed to bear the whole weight of the body. They include llamas, camels, antelopes, deer, oxen, pigs and hippopotamuses (*see* PERISSODACTYLA).

ascorbic acid another term for the water-soluble vitamin C found in all citrus fruits and green vegetables (especially peppers). A deficiency of ascorbic acid leads to the fragility of tendons, blood vessels and skin, all of which are characteristic of the disease called scurvy. The presence of ascorbic acid is also believed to help in the uptake of iron during the process of digestion by the body.

asexual reproduction new individual formed from a single parent without the involvement of any sexual process or production of gametes. It occurs in plants, micro-organisms and some lower animals and may alternate with a sexual phase. Different methods are employed e.g. in amoeba by FISSION, and by FRAGMENTATION in some annelid worms. In coelenterates another method that is employed is BUDDING. In plants, the most familiar means is by the production of bulbs, corms and tubers (VEGETATIVE PROPAGATION). Ferns, mosses and some other plants show an alternation of generations with the asexual stage being the production of spores—the SPOROPHYTE generation.

aspirin *or* **acetyl salicylic acid** a drug widely used to relieve pain and reduce fever and INFLAMMATION. The administration of aspirin inhibits the PROSTAGLANDIN production which causes inflammation. Aspirin also acts to reduce the joining together of BLOOD platelets, hence helping to maintain blood flow, and is used to treat circulatory disorders.

atmosphere (1) a unit of pressure defined as the pressure that will support a mercury column 760mm high at 0°C, sea level, and a latitude of 45°. (2) the layer of gases surrounding the earth, which contains, on average, 78 per cent nitrogen, 21 per cent oxygen, almost 1 per cent argon, and then very small quantities of carbon dioxide, neon, helium, krypton and xenon. In addition, air usually contains water vapour, hydrocarbons and traces of other materials and compounds.

atom the smallest particle that makes up all matter and still retains the chemical properties of the element. Atoms consist of a minute nucleus containing PROTONS (p) and NEUTRONS (n), with negatively charged particles called ELECTRONS (e) moving around the nucleus. The number of protons in an atom is equal to the number of electrons, and as the protons are positively charged, the atom is, overall, electrically neutral (*compare* ISOTOPE). The various elements of the PERIODIC TABLE all have a unique number of protons within their atomic nucleus.

atomic mass unit (*see* RELATIVE ATOMIC MASS) defined in 1961 as one twelfth of the mass of an atom of ^{12}C, having formerly been based upon ^{16}O, the most abundant ISOTOPE of oxygen.

atomic number (A, at. no.) the number of protons in the NUCLEUS of an ATOM. Although all atoms of the same element will have the same number of protons, they can differ in their number of neutrons, resulting in an ISOTOPE.

atomic weight *see* **relative atomic mass**.

ATP (*abbreviation for* adenosine triphosphate) an important molecule that is used as energy to drive all cellular processes. It consists of ADENINE and a 5-ring sugar that has three phosphate (PO_4) groups attached by high-energy bonds. ATP can be synthesized during GLYCOLYSIS by addition of a phosphate group to adenosine diphosphate (ADP), or it can be broken down to form the ADP, releasing energy that will be used to drive a metabolic process, such as active transport across cell membranes or the contraction of muscle cells.

atrium (*plural* **atria**) a minor chamber of the heart that is considered to be a reservoir as blood passes from it into the pumping chamber, the VENTRICLE. The right atrium of the heart receives the blood carried by the superior and inferior VENA CAVA before it passes via a valve into the right ventricle. The pulmonary veins carry oxygenated blood from the lungs into the left atrium, which then flows via a valve into the left ventricle.

atrophy the process whereby an organ or part of the body degenerates and reduces in size, usually involving destruction of cells. It may be controlled hormonally and genetically and can also result from prolonged starvation.

atropine a poisonous ALKALOID of the tropane group with a crystalline structure. It occurs naturally in the family of flowering plants known as the Solanaceae, which includes *Atropa belladonna*— deadly nightshade. Used medically it has a dampening down effect on the central nervous sytem and acts on the VAGUS NERVE to decrease secretions in the respiratory tract and increase the heart beat rate. It also causes the pupil of the eye to dilate and can be used in the treatment of colic.

autoclave equipment used to carry out chemical reactions at high temperatures and pressures. Also for sterilization procedures using steam under high pressures and temperatures. After 12 minutes of being heated at 121°C and 138–172 kNm pressure, bacterial spores, viruses and vegetative material are destroyed.

autoimmunity a complex condition, which is not entirely understood, in which the IMMUNE SYSTEM attacks its own cells or cell components. This breakdown of the body's normal defence mechanism is a cause of several debilitating and potentially fatal human disorders such as rheumatoid arthritis and rheumatic fever.

autolysis *see* **lysis**.

autonomic nervous system (ANS) the part of the vertebrate nervous system that supplies the body's

SMOOTH MUSCLE (e.g. the gut) cardiac (heart) muscle and glands with a motor nerve supply. It is sometimes called "involuntary" because its activity regulates the internal environment of the body, and two divisions are recognized, the SYMPATHETIC and PARASYMPATHETIC NERVOUS SYSTEMS.

autosome a biological term describing all CHROMOSOMES within a cell except the SEX CHROMOSOMES. In a DIPLOID cell, there are two copies of every autosome, each of which will carry genetic information for the same aspect of the individual's PHENOTYPE. Although autosomes are not involved with determining the sex of an individual, they can carry genetic information that will affect the sexual characteristics.

autotrophism self-feeding—when an organism manufactures the organic substances it requires to sustain life from inorganic sources. Green plants are autotrophs, using carbon dioxide derived from the air and minerals and water in the soil, and light as an energy source. Hence they are photoautotrophs. More rarely, some bacteria are chemoautotrophs and derive their energy from chemical reactions using such inorganic substances as ammonia and sulphur.

auxin any one of a number of substances that act to promote aspects of plant growth. The most familiar naturally occurring auxin is indoleacetic acid (IAA) which is manufactured in the tips of shoots. Auxin affects plant growth in several ways but one of the most important is to cause the enlargement of cells in developing, young shoots (*see also* GROWTH SUBSTANCE). Synthetic auxins, e.g. 2,4,5-T and 2,4-D, have been widely, and sometimes controversially, used as weedkillers.

Aves the class of vertebrates that comprises the
birds. Birds have a reptilian ancestry and evolved at
some stage during the Mesozoic era (*see* APPENDIX 5).
They retain some features of their reptilian origins
such as the production of amniote eggs and scales on
the legs. The most familiar fossil representative is
Archaeopteryx, which dates from the Upper Jurassic.
Two modern superorders exist, the Palaeognathae
(ratites) and Neognathae (containing the largest
order, the Passenformes or perching birds, which
includes the songbirds). The body of a bird is typically highly modified for flight with a lighter skeleton than that present in other vertebrates. Some
organs are absent to reduce weight, e.g. only one
ovary in females. The front limbs are developed as
wings which are aerodynamically adapted for flight.
Feathers, composed of KERATIN, cover the body and
those on the wings are highly modified for flight
while others provide insulation. The breastbone is
well-developed, having a KEEL for the attachment of
large flight muscles. Birds do not have teeth and
food is ground up in a GIZZARD. The jaws are developed to form the beak or bill, also made of keratin,
which varies enormously in size and shape between
the different species. The adaptations of the beak
enable the birds to exploit a great number of different habitats. Birds have excellent eyesight and
highly developed social behaviour. Internal fertilization takes place and hard-shelled eggs in which
the young develop are produced and incubated.

axon *see* **neuron**.

B

Bacillariophyta the diatoms, a division of the algae. Both marine and freshwater forms occur and they are major components of plankton and important in the food chain. They are single-celled plants but sometimes form chains or colonies. The cell walls are unique, being glassy, made of SILICA embedded in PECTIN, and in two halves, one overlapping the other. There are slits in the walls and the pattern of these pores and overall shape of the cells is used to distinguish different species. Reproduction is mainly asexual with cell division by MITOSIS. Half of the cell wall goes to each daughter cell which then regenerates the other half. Diatoms are very important fossils, and enormous accumulations of their cell walls are mined as diatomaceous earth (kieselguhr), which has a variety of uses. Food is stored as oil and this has contributed to the petroleum reserves that are exploited today.

bacillus (*plural* **bacilli**) a general term for any rod-shaped bacterium. Also a genus of bacterium found in soil and the air.

backcross a mating experiment that is used to discover the GENOTYPE of an organism. It involves crossing the organism of unknown genotype with an organism of known genotype (usually the HOMOZYGOTE recessive). The PHENOTYPES of the produced progeny should directly correspond to the chromosomes of the parental organism of unknown genotype. A backcross usually reveals whether the unknown genotype is homozygous or heterozygous (*see* HETEROZYGOTE) for a particular gene.

bacteria (*singular* **bacterium**) PROCARYOTE, mainly single-celled organisms belonging to the kingdom Monera. They are among the simplest of organisms and were the first to evolve, but their impact upon life on Earth is vast as they underpin all life-sustaining processes. A slimy, protective outer layer called a capsule may be present and also filaments and FLAGELLA which effect movement. A test known as GRAM'S STAIN is used to distinguish two types of bacteria which have a different cell wall structure (Gram positive and Gram negative). They are also distinguished according to shape and there are many different forms: spiral (spirilli), spherical (cocci), rod-like (bacilli), comma-shaped (vibrio) and corkscrew-like (spirochaetae). They are also described according to the conditions in which they exist, e.g. AEROBIC or ANAEROBIC. Bacteria are the vital means by which organic material is decomposed and substances made available for plant growth. They are the main agents in the chemical cycles of CARBON, oxygen, NITROGEN and sulphur. Most are saprophytes or PARASITES and some are autotrophs (*see* AUTOTROPHISM). Reproduction is largely asexual by simple fission but some have a form of sexual reproduction. Others produce highly resistant spores. A few are responsible for diseases in plants and animals, e.g. cholera, diphtheria, syphilis, tuberculosis and typhoid.

bacteriophage *or* **phage** a virus that infects a BACTERIUM. Being specific to a particular bacterium, phages find use in genetic engineering as a vehicle in cloning (*see* GENETIC ENGINEERING) and in certain manufacturing processes that utilize bacteria, e.g. the production of cheese.

baleen *see* **whalebone**.

barb in plants it is a hooked hair and in animals a bristle-like or hooked structure. The barbs of a feather project sideways from the central shaft (the rachis) and in turn give rise to filaments known as barbules (*see* FEATHER). These interlock to form the surface or vane of the feather.

bark a general term for the outer protective layer of woody stems and roots composed mostly of cork. Sometimes bark is composed of alternating layers of cork, dead PHLOEM cells and dead CORTEX and this is known as a *rhytidome*.

barrier reef a reef of coral built up parallel to the shore but some way from it, so as to create a lagoon. A good example is the Great Barrier Reef, Australia, which is almost 2000 kilometres (1250 miles) long.

base in a chemical reaction, any substance that dissociates in water to produce hydroxide ions (OH^-). A base can be thought of as a proton (H^+) acceptor or as an electron pair donor. It will react with an acid to give a salt and water (the latter formed from the OH^- ion from the base and the H^+ ion from the acid).

$$NaOH(aq) + HCl(aq) \longrightarrow NaCl(aq) + H_2O$$

base pair the arrangement of two nitrogenous molecules on opposite strands of a DNA double-helix or a DNA-RNA molecule. The bases can be classified according to their structure: ADENINE and GUANINE are PURINES, whereas THYMINE, CYTOSINE and URACIL are PYRIMIDINES. As a consequence of geometrical factors, a purine will be hydrogen-bonded to a pyrimidine, i.e. A-T and C-G. This specific base-pairing keeps the DNA structure in a highly organized order, allowing the replication of DNA to be very precise, thus ensuring that each daughter cell will inherit the same genetic information contained within the parent cell.

B-cells these are LYMPHOCYTES, which differentiate in the bone marrow to form part of the IMMUNE SYSTEM of humans. B-cells become activated when they encounter a specific ANTIGEN, leading to proliferation and secretion of ANTIBODIES by the activated B-cells. After a first encounter with an antigen, some of the B-cells remain and serve as memory cells. The memory cells will be capable of recognizing the same antigen during any subsequent encounter and will, therefore, produce a faster and greater secondary response (this is the principle behind vaccination).

becquerel the unit of radioactivity in the SI scheme, defined as the activity of a radionuclide decaying at one nuclear transition per second.

Benedict's test devised by the American chemist S. R. Benedict (1884–1936) to test for GLUCOSE and for reducing sugars. It is a modification of FEHLING'S TEST. It consists of a solution of copper sulphate, sodium carbonate and sodium citrate present in certain proportions. A small amount is added to the test solution and boiled. Any sugar present is oxidized and a precipitate of rust-coloured cuprous oxide is formed. This test is used medically to detect the presence of sugar in urine if DIABETES is suspected.

benthic plant or animal living on the bottom of a lake or sea. This may include crawling or burrowing at the sediment-water boundary, attached to it (as with seaweeds) or purely sessile.

benzene a toxic hydrocarbon that is a liquid at room temperature and has the chemical formula C_6H_6. The six carbon atoms of benzene form a ring, and the overall molecular structure is planar, with all bond angles having the same value of 120°. The benzene ring is a very stable structure due to the delocalization of six electrons (one electron is contributed by each carbon atom). All AROMATIC HYDROCARBONS contain a benzene ring within their molecular structures.

beri-beri a crippling human disease that is caused by dietary deficiency of the water-soluble vitamin B, also called THIAMINE, which is essential for the metabolic conversion of carbohydrate to glucose. Affected individuals show symptoms of muscle atrophy, paralysis, and mental confusion, and may eventually suffer from heart failure. Chronic alcoholism will lead to beri-beri, and this is thought to be responsible for the development of a psychosis called KORSAKOFF'S SYNDROME.

bilateral symmetry the case when an organism can be "split" into two halves which are nearly mirror images of each other.

bile a viscous fluid produced by the liver and stored in a small organ, the gall bladder, near the liver. Bile is an alkaline solution consisting of bile salts, bile pigments and CHOLESTEROL, which aids in the digestion of fatty particles present in the diet. Food

entering the DUODENUM triggers the muscular contraction of the gall bladder wall, and bile is forced
into the duodenum via the bile duct. Although bile
does not contain any digestive ENZYMEs, the bile
salts help in the digestion of fatty food particles by
acting as emulsifying agents, i.e. they break down
large fat particles into many smaller ones, a process
that exposes a larger surface area to the digestive
action of the enzymes, LIPASES.

binomics *see* **ecology**.

biochemistry a method for investigating the chemistry of the biological processes occurring in the cells
of all organisms. Such investigations provide an
understanding of a broad range of important processes, from the control of cell metabolism to the
effects that a disease has upon the cells of the body.

biological control human interference to control a
pest or parasite by biological (and ecological) means
rather than by using chemicals. The attempted
control can be exercised in different ways, perhaps
by exploiting a predator-prey relationship, e.g. the
introduction of fish to consume mosquito larvae.
The spread of the prickly pear cactus in Australia
was successfully controlled by the introduction of
the cactus moth, whose caterpillars fed on and
destroyed the young shoots. PHEROMONES are now
used to attract certain insect pests that may then be
destroyed, or sometimes the males rendered infertile by radiation and then released. Infertile matings
then cause a drop in overall numbers. Although the
technique can confer a number of advantages and
has been used successfully, great care must be
exercised and the biological implications thoroughly
understood.

biological (*or* **biochemical**) **oxygen demand** (BOD) a measure of the pollution of effluent where micro-organisms take up dissolved oxygen in decomposing the organic material present in the effluent. BOD is quantified as the amount of oxygen, in milligrams, used by one litre of sample that is stored in the dark at 20°C for five days.

bioluminescence (*see also* LUMINESCENCE) the production of light by living organisms such as some bacteria, fungi, deep-sea fish, coelenterates, crustaceans, fireflies and glow-worms. There may be special sites and organs for the production of bioluminescence or organisms may utilize "photogenic" bacteria in a symbiotic relationship (*see* SYMBIOSIS). The light may be emitted continuously, as in bacteria, or intermittently, as in fireflies. It serves a number of functions such as mate or prey attraction, and protection. The photoprotein involved in light production is called luciferin and its composition varies according to the species. It is catalysed by the ENZYME luciferase using ATP, and oxidized in the presence of oxygen during which reaction light is released as luciferin returns to its stable condition.

biome a geographical area of the Earth that is characterized by particular ecosystems and animal or plant communities and is related to (and created by) climatic pattern. The dominant form of vegetation usually distinguishes the biome, e.g. tundra, but animals, micro-organisms, etc, also typify a specific biome. The major terrestrial biomes are desert, savanna, tropical forest, chaparral (scrub land), temperate grassland, temperate DECIDUOUS forest, taiga (boreal or coniferous forests) and TUNDRA.

biosphere the region of the Earth's surface (both land and water) and its immediate atmosphere, which can support any living organism.

biosynthesis the production of complex chemical compounds by living organisms using ENZYMES.

biotechnology the industrial use of organisms, their parts or processes, to produce drugs, food or other useful products. Modern processes include the controlled growth of specific fungi in laboratories to obtain the antibiotic penicillin, and the production of alcohol during FERMENTATION of yeast. The scope for biotechnology is enormous, with a great deal of research being directed towards GENETIC ENGINEERING in plants and animals.

biotic relating to life or living things (hence biota, the animal and plant life of a region). Thus for an organism, the other living organisms around it comprise the biotic environment and may be competitors, predators, parasites, etc.

biotin one of the vitamins in the B-complex that is made by bacteria present in the animal intestine. It is also found in milk and liver, vegetables and cereals. A deficiency of biotin can only occur if large amounts of raw egg white are ingested as this contains a protein, avidin, which binds to biotin. A modified form of biotin acts as a co-enzyme operating in part of the KREBS CYCLE.

biuret test a biochemical test to detect the presence of PROTEINS and PEPTIDE BONDS (but not amino acids) in a solution. Biuret itself ($H_2NCONHCONH_2$) is formed when urea is heated and so the test can be used medically to detect this. The test sample is mixed with sodium hydroxide solution and then a small amount of 1 per cent copper sulphate solution is slowly added. A positive result is indicated by the production of a purple ring where the two solutions meet.

Bivalvia a class of the phylum Mollusca, also known as **Pelecypoda**, containing many species, e.g. musssels, oysters, scallops, clams and razorshells. Both freshwater and marine forms occur and they have a body that is laterally flattened and enclosed within a shell divided into two halves (bivalve shell). The shell is hinged along the mid-dorsal line and drawn tightly together by powerful adductor muscles. A siphon draws in water which passes over two large GILLS where both exchange of gases and filter feeding occurs. The used water passes out through another siphon. Bivalves have a greatly reduced head compared to other molluscs, and tend to be sedentary, living on the sea or lake floor, although burrowing forms (shipworm, razorshell) and more mobile ones such as scallops do occur.

bladder in animals it describes any sac-like structure in which liquid or gas is retained, e.g. the gall bladder (see BILE) in humans and the SWIM BLADDER in fish. It more usually refers to the urinary bladder present in most vertebrates; a muscular sac in which URINE collects before being expelled to the outside. In mammals the urine is produced in the paired KIDNEYS and passes to the bladder through two tubes called URETERS. A single tube, the URETHRA, leads from the bladder to the outside. In some aquatic, insectivorous plants, such as the bladderwort, the bladder is a ball-shaped hollow with a single opening controlled by a valve. It acts as a trap that catches the prey on which the plant feeds. In some seaweeds, the bladderwrack, for example, it is an air-filled sac that provides buoyancy.

blastula in animals this is formed during early embryonic development by the cleavage of a fertilized egg (OVUM). It usually resembles a hollow ball in which the dividing cells (blastomeres) of the early embryo form a layer (the blastoderm) that surrounds a cavity (the blastocoel). There is no blastula in the development of insect eggs. The blastula forms a disc (the blastodisc) on the surface of the yolk in vertebrates, and in mammals this stage is known as the blastocyst. It is succeeded by the GASTRULA.

blood a vital substance consisting of red blood cells (ERYTHROCYTES) and white blood cells (LEUCOCYTES) suspended in a liquid medium called BLOOD PLASMA. As well as many proteins, mammalian blood also contains small disc-shaped cells called platelets, which are involved in BLOOD CLOTTING. Blood circulates throughout the body and serves as a mechanism for transporting many substances. Some of the essential functions of blood are:

(1) Oxygenated blood is carried from the heart to all tissues by the arteries while the veins carry deoxygenated blood, which contains carbon dioxide, back to the heart.

(2) Essential nutrients, such as glucose, fats and AMINO ACIDs (the building blocks of proteins), enter the blood from the intestinal wall, or the liver and fatty deposits, and are carried to all the regions of the body.

(3) The metabolic waste products, ammonia and carbon dioxide, are carried to the liver, where they react to form urea, which is then carried by the blood to the kidneys for excretion.

(4) Steroid and thyroid hormones—important regulatory molecules—are carried to their target cells after they are secreted into the blood by the ENDOCRINE SYSTEM. Although insoluble molecules, they are carried in soluble particles called low density LIPOPROTEINS.

blood clotting (*also called* **haemostasis**) a process, involving many chemical factors, that stops blood leaking from an area of injured tissue. In the first instance, constriction of any blood vessels in the injured tissue restricts the leakage of blood, and the subsequent formation of a plug helps to seal off the damaged area, preventing the entry of micro-organisms. The formation of this plug is triggered by an enzyme secreted by damaged blood vessels and blood platelets and is completed after a chain of chemical reactions. The final hard clot consists of blood platelets, trapped red blood cells, and fibrin (*see* FIBRINOGEN).

blood grouping a method for classifying blood types by checking which particular ANTIGENs are present on the surface of red blood cells. More than four hundred antigens can be recognized by their specificity for a particular ANTIBODY. There are many systems for the classification of blood types, but the ABO blood group and rhesus (Rh) group are two important systems that are widely known. In the ABO blood group system there are basically two antigens, designated A and B. The A and B antigen may be present singly or together (AB), and the absence of both antigens gives rise to blood group O. There are thus four blood groups—A, B, AB or O. Naturally occurring antibodies to the ABO system develop only after the age of three months. A person blood plasma

of blood group A will have antibody-B present in their serum, i.e. anti-B serum. A person of blood group B has anti-A in their serum. A person of blood group O has both anti-A and anti-B, whilst a person of blood group AB has no antibodies.

In blood transfusions, it is vital that blood groups are correctly matched since incompatibility will result in blood clotting, which could cause the recipient's death. For instance, if the donor is blood group A and the recipient is blood group B, then anti-B serum of the donor will react with the B-antigen present in the recipient's blood and initiate blood clotting. Of course, the A-antigen of the donor's blood will also react with the anti-A in the recipient's blood.

The rhesus blood group system can be simply explained in terms of whether an individual is Rh-positive or Rh-negative. The presence of an Rh-factor (D-antigen) on the surface of red blood cells will classify an individual as Rh-positive, whereas the absence of such a factor will classify the individual as Rh-negative.

blood plasma blood from which all the blood cells (ERYTHROCYTES, LEUCOCYTES and platelets) have been removed. The resulting solution is 90 per cent water and contains some proteins, sugar, salt, urea, hormones and vitamins.

blood serum a fraction of the liquid medium of blood, i.e. plasma, minus one of the plasma proteins, called FIBRINOGEN.

blood type *see* **blood grouping**.

blue-green algae *see* **cyanobacteria**.

BOD *see* **biological oxygen demand**.

Bohr effect the discovery by the Danish physiologist

Christian Bohr (1855-1911) that the oxygen-carrying capacity of blood varies within different parts of the body. He showed that the pH of the body tissue determined whether oxygen would be released from, or taken up by, the HAEMOGLOBIN in red blood cells. At low pH (acid conditions), oxygen is released from haemoglobin and enters the surrounding tissues, but at high pH (alkali conditions), oxygen from the surrounding tissue will bind strongly to the haemoglobin in the red blood cells. The Bohr effect explains why oxygen is taken up by the haemoglobin in blood circulating throughout lung tissue (high pH) but is released by blood circulating in active muscle sites (low pH).

boiling point the temperature at which a substance changes from the liquid state to the gaseous state. It occurs when the vapour pressure of a liquid surface equals the surrounding atmospheric pressure. The boiling point of a pure liquid is measured under the standard atmospheric pressure of 1 atmosphere (equivalent to 760mm mercury).

bond the force that holds atoms together to form a MOLECULE, a compound or a lattice. The type of bond between neighbouring atoms will be determined by the electron attraction strength of each atom and can be of three types—COVALENT, IONIC or POLAR COVALENT.

bone the matter that forms the skeleton of most vertebrates, it is a hard connective tissue comprising a matrix of 30 per cent COLLAGEN fibres and 70 per cent bone salts—crystalline calcium phosphate (hydroxyapatite) in which are embedded the bone cells (oestoblasts and oestocysts). The bone cells are responsible for the formation of the matrix. There are two types of bone, compact and spongy. Compact bone has a series of concentric layers, the lamellae, that surround the calcified matrix, the bone cells and also small canals known as HAVERSIAN canals. It forms the cylindrical shafts of the vertebrate long bones. Spongy bone occurs on the inside of the long bones and also at the ends. It forms most of the flat bones such as those of the skull and also the vertebrae. It has a similar composition but contains trabeculae (bars of tissue) which form an interlocking network. Embryonic cartilage is generally replaced in the growing animal by bone, which is well supplied with blood vessels and nerves located within the Haversian canals (*see also* ENDOSKELETON).

bone marrow a spongy soft tissue that occupies the central space of the long bones and also occurs within the internal cavities of other bones. In young animals all bone marrow is responsible for the production of blood cells and is known as red marrow. (In older animals the marrow of the long bones is replaced by fat, known as yellow marrow, and does not produce blood cells. Hence in mature animals red marrow occurs principally in such bones as the vertebrae, pelvis, ribs and breastbone). It contains MYELOID TISSUE in which occur ERYTHROBLASTS that differentiate into red blood cells (ERYTHROCYTES). Myeloid cells also develop into the LEUCOCYTES of the blood, which further differentiate into other cell types.

bony fishes (class Osteichthyes) have a body covered with overlapping scales and a bony skeleton containing calcium phosphate (*see* CHONDRICHTHYES). Marine and freshwater species occur, and respiration is through GILLS covered by a protective flap, the

OPERCULUM. Typically, a slimy mucus is secreted by glands in the skin which helps to reduce drag during swimming. Often present is an air sac, the SWIM BLADDER, which acts as a buoyancy aid. Jaws are present (*compare* AGNATHA) and usually external fertilization takes place with large numbers of eggs being laid.

botulism the most dangerous type of food poisoning in the world, caused by the anaerobic bacterium called *Clostridium*. This bacterium is found in an oxygen-free environment, such as underneath soil or in an airtight food can. During growth, it releases toxins that, if bacteria are living in the cells of the body, will affect the nervous system of humans. This can result in death, especially if it is the vulnerable members of a population who are affected, e.g. newborn babies and elderly people.

bp *abbreviation for* BASE PAIR, BOILING POINT.

Brachiopoda an ancient, but now small, phylum of marine invertebrates, the lamp shells. Along with two other phyla these animals are called lophophorate, which describes their most distinctive characteristic, the LOPHOPHORE, a structure used in feeding and respiration. The brachiopods have a bivalve shell, the two halves of which are DORSAL and VENTRAL to the animal. A stalk (PEDUNCLE) attaches the animal to the surface on which it lives. Brachiopods are very important INDEX FOSSILS. One genus, *Lingula* has survived to the present day little changed from the Ordovician, 500–440 million years ago.

brackish water which is intermediate in saltiness, between the sea and fresh water, as occurs in estuaries (*see* ESTUARY).

brain a general term describing the main ganglionic

mass in an animal's CENTRAL NERVOUS SYSTEM. In those invertebrates where a brain can be recognized, the term usually refers to paired ganglia at the anterior end. However, similar activities to those carried out in the "brain" may occur in different pairs of ganglia in these animals.

In vertebrates, the brain is a highly complex and enlarged part of the central nervous system, situated at the anterior end of the body. It is connected with the rest of the central nervous system by means of the SPINAL CORD. The brain is responsible for interpreting all the information received from an animal's sense organs (*see* NERVE IMPULSE), and sending out electrical signals to control muscles. It can be said to control nearly all bodily functions and is divided into a number of different regions. Internal cavities are present (*ventricals*) which are filled with CEREBROSPINAL FLUID bathing all the brain. Three membranes suround the brain and these are called the MENINGES. In vertebrate embryos the brain develops from three expansions of the NEURAL TUBE to form the FOREBRAIN, MIDBRAIN and HIND BRAIN. These develop further to form highly specialized regions performing different functions.

brain death occurs when there is a complete and continuous absence of the reflex and respiratory activities that are normally controlled by the MIDBRAIN area of the BRAINSTEM. An ELECTROENCEPHALOGRAM will show a flat trace but brain death is not necessarily associated with an absence of heartbeat. In humans, when the certainty of brain death is established, transplant organs may be removed.

brainstem a part of the vertebrate brain that resembles, and is a continuation of, the spinal cord and consists of the PONS, MEDULLA OBLONGATA and MIDBRAIN. It is the link between the spinal cord and pons and contains some of the RETICULAR FORMATION.

bronchus (*plural* **bronchi**) one of two tubular off-shoots from the trachea which then splits into smaller bronchi and then bronchioles. The bronchi walls are supported by cartilage rings. The bronchioles are very fine tubes that end in ALVEOLI, where carbon dioxide and oxygen are exchanged.

brown coal *see* **lignite**.

brown fat *see* **adipose tissue**

Brownian motion a phenomenon first discovered in 1827 by the Scottish botanist Robert Brown (1773-1858). He observed the random movement of minute particles that occurs in both gases and liquids. Brownian motion is taken as evidence to support the theory that KINETIC ENERGY is an inherent quality of all matter, as it is assumed that motion that can be seen also occurs in other substances where it cannot be seen.

Bryophyta commonly refers to a division of the plant kingdom containing simple plants belonging to such classes as the hornworts (Anthoceropsida), liverworts (Hepaticopsida) and mosses (Bryopsida). However, a modern view is that these three properly belong to separate divisions as, although they share many common features, they are probably not closely related. In this case only the mosses are contained within the division Bryophyta. These plants commonly thrive in wet, shady conditions but possess a waxy outer layer (CUTICLE) that helps to prevent them from drying out. They show an alternation of generations and in the sexual stage (GAMETOPHYTE generation) many male gametes or sperm are produced by a single sex organ known as an antheridium (*plural* antheridia). The sperm bear FLAGELLA and swim to the female gametes to fertilize them. Each female sex organ, the archegonium (*plural* archegonia) produces a single egg cell. Hence water must be present for fertilization to be effective, although in some plants a film of dew or rainwater on the surface may be sufficient. The sexual stage is followed by the SPOROPHYTE generation—a capsule on a stalk (seta) containing spores. The leafy, green gametophyte is the most important stage and the sporophyte is partly dependent upon it in order to obtain nutrients and water. In contrast to higher plants, VASCULAR TISSUE is absent or poorly developed. Also, there are no true roots although plants may be attached to a surface by rhizoids (hair-like outgrowths) or use other plants for support.

Bryozoa *see* **Ectoprocta**.

bud usually refers to the compact, immature plant shoot that is composed of numerous unexpanded leaves or flower petals that are rolled, folded or wrapped around one another. In leaf buds, the outer leaves may be hardened to protect soft, delicate inner leaves, and buds are typically borne on a short stalk. (The tip of a stem bears a terminal or apical bud, and an axillary or lateral bud occurs at axils, i.e. where a leaf meets the stem). The term also describes an outgrowth from a parent organism that becomes detached and grows into a new individual (*see* BUDDING).

budding a method of asexual reproduction in simple animals such as coelenterates (*see* CNIDARIA) and sponges (PORIFERA) where the parent organism produces an outgrowth that breaks free and develops into a new, genetically identical individual (also called gemmation). In plants, budding is found in YEASTS and other single-celled FUNGI. In horticulture, bud grafting of woody plants is commonly employed, especially in fruit trees and roses. A bud or shoot called a scion (*see* GRAFT), together with some of the stem beneath, is inserted into a host plant, the stock, by means of a cut or GRAFT, usually through the bark of the latter. The stock may be a closely related or wild species and stock and graft function as one plant once the procedure has been carried out.

buffer a chemical substance capable of maintaining the pH of a solution at a fairly constant value. It does this by removing hydrogen ions (H^+) from the solution when small amounts of an acid are added, or releasing hydrogen ions when small amounts of base are added. Most buffers are ionic compounds, usually the salt of a weak acid or base.

bulb an organ of vegetative reproduction that occurs underground and is formed from a shortened flattened stem and scaly, fleshy leaves. On the outside, dry, papery leaves occur. A terminal bud is produced in the centre and during the growing season it develops using food stored in the surrounding scale leaves which then wither. The bud produces a flower and leaves which photosynthesize (*see* PHOTO-SYNTHESIS) and pass food down into the leaf bases which then form a new bud and bulb at the base. More than one bud (and hence bulb) may be formed and this is known as VEGETATIVE PROPAGATION. Examples are tulips, daffodils, onions (*compare* CORM).

C

caecum (*plural* **caeca**) any small bag-like structure, usually of the gut, in animals, but particularly a blind pouch in the vertebrate ALIMENTARY CANAL, between the SMALL INTESTINE and COLON. The caecum shows the greatest degree of development in herbivorous mammals and houses the bacteria that digest cellulose. Sometimes a valve is present which increases the surface area available.

caffeine a PURINE that occurs in tea leaves, coffee beans and other plants. It acts as a weak stimulant to the central nervous system.

calcium a metallic element with the symbol Ca (*see* APPENDIXES 1 and 2). It is an essential constituent for normal growth and development in animals found in blood, teeth and bones. Its presence is essential for various metabolic processes, including muscle contraction. It is present in the middle lamella (a layer that binds together plant cells) in the form of calcium pectate.

calorie a unit of quantity of heat defined as that heat required to raise the temperature of one gram of water through 1°C. The calorie has in the main been replaced as a unit by the JOULE (1 calorie = 4.186 joules).

calorific value the number of heat units that are

obtained when a unit mass of a given substance undergoes complete combustion. In biology it is usually used to express the energy value of foods, i.e. when a food is oxidised within an animal body it is the amount of energy produced. Generally, this is expressed as kilojoules per gram (kJg^{-1}).

Calvin cycle the last stage of PHOTOSYNTHESIS, named after the American biochemist Melvin Calvin (1911–), who discovered that radioactive carbon dioxide ($^{14}CO_2$) became incorporated into the carbohydrate subsequently found in the cells of plants. All the various chemical reactions involved in the Calvin cycle occur in the stroma of plant cells. These reactions are termed the DARK REACTIONS of photosynthesis, since the formation of glucose can occur without light, although it does need the products generated by the light reactions, e.g. ATP. The fixation of carbon dioxide and its conversion into carbohydrate is an energy-requiring process, as it takes six turns of the Calvin cycle to produce one molecule of glucose. The required energy is generated by the HYDROLYSIS of ATP.

Cambrian the earliest geological period of the Palaeozoic era, dating from about 600–470 million years ago. Cambrian rocks contain the earliest known, abundant, animal fossils, which have all been marine animals, many with mineralized shells. Included are the GRAPTOLITES, brachiopods (*see* BRACHIOPODA), some worms, molluscs, echinoderms (*see* ECHINODERMATA) crustaceans, king crabs, eurypterids and TRILOBITES.

cancer a disease characterized by an uncontrolled growth rate of cells, leading to the formation of tumours. If the tumour remains localized, it is termed benign and is usually harmless to the host. However, malignant tumours do not remain localized but, instead, spread throughout the body and set up a secondary growth area by a process called METASTASIS. This usually causes death in the host as the malignant tumour disrupts the essential everyday processes of the cells in the affected tissues. There is not a single cause of cancer—it is triggered by a combination of factors, including exposure to carcinogens (such as tobacco smoke), radiation, ultraviolet light, certain viruses, and the possible presence of potential cancer genes (oncogenes). Treatment of different cancers involves surgery, radiotherapy, chemotherapy and, just recently in 1987, the "magic bullet" approach, where cytotoxic drugs are labelled with one specific ANTIBODY, which will only recognize a protein present on the surface of cancer cells.

canine teeth situated one on either side of upper and lower jaws between the first PREMOLARS and second INCISORS. Commonly known as dog or eye teeth, they are conical in shape, pointed and sharp and show the greater development in the CARNIVORES—wolves etc. They are used for piercing and tearing the flesh of prey and may be greatly elongated as in the tusks of wild boar. In herbivorous animals they may be reduced or even missing, as in the rabbit and giraffe.

capillary (*plural* **capillaries**) any narrow or hairlike tube with a thin wall usually only one cell thick. Capillaries form a network along which substances are transported, and through the thin walls of which materials are exchanged between themselves and their surrounding tissues. Blood capillaries exchange oxygen and nutrients through their walls and carry blood from the ARTERIOLES to all the body's living tissues. Metabolic waste products such as urea from the LIVER and carbon dioxide are also transported to the VENULES (and eventually excreted). Capillary walls lack connective tissue or muscle but can be dilated or constricted by means of precapillary SPHINCTERS which increase or restrict the blood flow in response to particular requirements. This regulation occurs in response to the presence of ADRENALINE and NORADRENALINE, changes in pH, temperature and amounts of oxygen and carbon dioxide. Capillaries in the LYMPHATIC SYSTEM are closed at one end with valves along their length, which drain excess fluid from tissue. BILE capillaries act in the transport of bile manufactured in the vertebrate liver.

capillary action a phenomenon related to SURFACE TENSION in liquids and due to inter-molecular attraction at the boundary of a liquid, which results in liquid rising or falling in a narrow tube.

capillary pressure (*see* CAPILLARY ACTION) for a capillary tube with radius r, and a liquid of surface tension γ with an angle of contact θ, the capillary pressure is defined as $P = 2\gamma \cos \theta / r$.

capsule a term widely used in the biological sciences.

(1) In flowering plants, it is a dry fruit, containing seeds, that is formed from several CARPELs adhering together. The seeds are released in various ways when the fruit is ripe—by splitting of the carpels (then known as valves), as in the iris; through a lid, as in the scarlet pimpernel; and through pores, as in the snapdragon.

(2) In mosses and liverworts (*see* BRYOPHYTA), a capsule containing spores borne on a thick stalk, the SETA, is produced—the SPOROPHYTE.

(3) Some BACTERIA produce a protective capsule of gelatinous material outside the cell wall.

(4) In animals, a capsule describes the connective tissue sheath that is present surrounding some vertebrate skeletal joints. Also it describes the fibrous or membranous covering surrounding certain vertebrate organs such as the KIDNEYs.

carbohydrates a large group of compounds containing carbon, hydrogen and oxygen, with the general formula $C_x(H_2O)_y$. The group includes the sugars, starch, and cellulose, and forms the mono-, oligo- and polysaccharides. Carbohydrates play a vital role in the metabolism of all living organisms.

carbolic acid *see* **phenol**.

carbon a non-metallic element, symbol C, that is present in all organic compounds and hence occurs in every living organism. Carbon atoms bind with each other and also with oxygen, nitrogen, hydrogen and sulphur atoms. Stable, COVALENT BONDING occurs and the compounds formed have a variety of configurations of carbon atoms in chains and rings. The carbon is made available to living organisms through atmospheric carbon dioxide, then taken up by plants during PHOTOSYNTHESIS (*see* CARBON CYCLE).

carbon compounds based upon carbon (C), element 6 in the PERIODIC TABLE, these include all organic compounds and are the basis of all living matter.

carbon cycle the circulation of carbon compounds in the natural world by various metabolic processes of many organisms. The main steps of the carbon cycle are:

(1) Carbon dioxide in air and water is taken up during photosynthesis in plants and some bacteria. (2) The carbon accumulated in plants is later released during the decomposition of the dead plant, or of bacteria or animals that have consumed any of the plant. (3) Carbon will also be released by the burning of fossilized plants in the form of fuels—coal, oil and gas—and during the respiration of all organisms. The concentration of carbon dioxide in the atmosphere is increasing as huge areas of tropical forests are destroyed, while the consumption of fossil fuels is rising, i.e. less PHOTOSYNTHESIS to absorb the increasing CO_2 level. This may be a factor involved in the small temperature rises throughout the world known as the GREENHOUSE EFFECT. When there are high levels of carbon dioxide in the atmosphere, heat radiation from the sun tends to be reflected back to earth rather than lost to space.

carbon dioxide (CO_2) a colourless gas occurring in the ATMOSPHERE due to OXIDATION of carbon and CARBON COMPOUNDS. It is the source of carbon for plants and plays a vital role in METABOLISM. It solidifies at -78.5°C and is much used as a refrigerant. It is also used in carbonated drinks and, since it does not support combustion, in fire extinguishers.

Carboniferous a geological period in the Palaeozoic era, following the Devonian Period and preceding the Permian Period, from 360–286 million years ago. In Europe, it is divided into the Lower and Upper Carboniferous. During Lower Carboniferous times the predominant environment was one of shallow seas in which the typical rock type, the carboniferous limestone, was deposited. This rock is rich in fossils, particularly invertebrates such as corals, brachiopods (*see* BRACHIOPODA) and bryozoans (*see* ECTOPROCTA). In Upper Carboniferous times, deltaic conditions gave way to an environment of swamp forests where giant lycopods (division Lycophyta) and enormous horsetails (division Sphenophyta) thrived. Also present were the ferns (division Pterophyta) and primitive seed plants that did not become predominant until the swamps dried out at the end of the Carboniferous period. The typical rocks of the Upper Carboniferous are the coal measures where beds of shale, sandstone, clay and coal occur as alternating layers. The coal was formed from the plant remains which did not decompose completely in the stagnant waters of the swamps and formed thick accumulations of peat. Through time more rocks were formed on top and heat and compression converted the peat to coal.

carbon monoxide (CO) a colourless gas formed during the incomplete combustion of coke and similar fuels. It also occurs in the exhaust fumes of motor engines. Carbon monoxide is poisonous when breathed in because it combines with the HAEMO-GLOBIN in the BLOOD to form a stable compound. This reduces the oxygen-carrying capacity of the blood. It is a valuable industrial reagent because of its reducing properties.

carboxyl the acid group -COOH, characteristic of the carboxylic acids where an oxygen is double-bonded and the hydroxyl (OH) is singly bonded to the carbon.

carboxylase one of a number of specific ENZYMES that operate in the fixing of carbon dioxide or the transferring of COO-groups. More specifically, it catalyses the decarboxylation of PYRUVIC ACID which is changed to ethanol. Carboxylases operate in the processes of PHOTOSYNTHESIS and RESPIRATION and are found in plants, yeasts and bacteria.

carboxylic acid an organic acid containing one or more CARBOXYL groups e.g. propanoic acid, CH_3CH_2COOH.

carcinogen a substance that may produce CANCER.

cardiac muscle a particular type of muscle that is found only in the heart wall of vertebrates and is highly specialized. Cardiac muscle fibres branch, rejoin and interlock and each has one nucleus. The operation of the muscle is normally involuntary, so does not require nervous stimulation, although activity in the VAGUS NERVE can affect heart beat rate (*see* PACEMAKER).

cardiovascular system the organization of the heart, the arteries and veins within the human body, in which the heart and the blood vessels form a virtually closed system. The minor branches of the blood vessels supply blood to every part of the body, including the bones. The only bloodless parts of the body are dead structures such as the nails and hair. However, the nail-bed and the hair roots do require a blood supply. The only other organ without its own blood supply is the cornea (the clear window of the eye). The cornea is supplied with oxygen and nutrients by means of the tears. Other than these exceptions, every part of the body requires a constant blood supply in order to receive essential nutrients such as GLUCOSE, AMINO ACIDS, etc. Any interruption in the blood supply of the cardiovascular system causes death of that tissue.

carina *see* **keel**.

carnassial teeth in CARNIVORES, the last premolars in the upper jaw and the first molars in the lower jaw which have cusps with sharp, cutting edges for slicing through flesh.

Carnivora the order containing mammals that are largely flesh-eating. Some are predators and some carrion eaters, while many combine the two to form a highly successful group. The construction of the jaw and its muscles confers a very powerful bite and the CANINE TEETH are well developed. Other cheek teeth may be further adapted for shearing flesh (*see* CARNASSIAL TEETH). Retractile claws are commonly present, and two sub-orders are recognized, the Fissipedia (toe-footed carnivores), e.g. wolves, dogs, foxes, cats, racoons, weasels and badgers, and the Pinnepedia (fin-footed carnivores), e.g. walruses, sea lions and seals.

carnivore any animal that eats the flesh of other animals. The term can also refer to the mammalian order, *Carnivora*, which includes bears, cats and dogs. CARNIVOROUS PLANTS are ones that trap and digest insects (using special ENZYMES) in order to obtain their nitrogen requirements.

carnivorous plants about 400 different species of plants that are usually insectivorous. They often grow in poor conditions which would not normally supply suffcient mineral nutrients. Hence they are specially adapted to attract, trap and digest mainly insect prey in ingenious and highly effective ways. For example, sundews secrete a sticky substance on the leaves which holds insects fast. Pitcher plants have modified leaves forming a water-filled trap with smooth sides into which an insect falls and drowns. Venus fly traps have hinged leaves with spines at the edges which snap shut over an insect. Once an insect (or other small prey) is caught, by whatever means, digestive ENZYMES that are secreted by the plant enable nutrients to eventually be assimilated.

carotenoids plant pigments, coloured orange, red and yellow, that chemically resemble TERPENES and are based upon the tetraterpene unit ($C_{40}H_{64}$). The group includes the carotenes, which are simple hydrocarbons, and the xanthophylls (oxygenated derivatives of carotenes). The characteristic colours of, for example, carrots and ripe tomatoes, are due to carotenoids. Although they are not essential to PHOTOSYNTHESIS they absorb photons and the energy is transferred to the CHLOROPHYLL. In addition these accessory pigments protect cells in some bacteria from photo-oxidation.

carotid artery one of a major pair of vertebrate arteries that carries oxygenated blood from the heart to the head. A pair of common carotid arteries branch off from the AORTA on the left of the heart and the inominate artery on the right. These run up on either side of the neck and branch into the internal and external carotids, which supply blood to the head.

carpel the female reproductive organ of a flower comprising stigma, style and ovary. The ovary holds one or more OVULES, which form seeds after fertilization.

carpus (*plural* **carpi**) in terrestrial vertebrates, this is the part of the forelimb corresponding to the wrist in man. It consists of a large number of small bones, the carpals, possibly 10–12 in some animals, but reduced to 8 in human beings. The carpals confer great flexibility at the wrist joint, and articulate with the RADIUS and ULNA of the arm and the metacarpals of the hand (*see* PENTADACTYL LIMB).

cartilage a type of connective tissue that, together with bone, forms the vertebrate SKELETON. It is composed of a stiff matrix of chondrin (mainly chondroitin sulphate, fine COLLAGEN fibres and GLYCOPROTEINS). This is secreted by cells called chondroblasts, which then come to occupy spaces (lacunae) within the matrix where they are known as chondrocytes. Nerves and blood vessels are absent from cartilage and its surface is surrounded by an irregular layer of connective tissue known as the perichondrium. Cartilaginous fishes (*see* CHONDRICHTHYES) have a skeleton composed entirely of cartilage, as does the vertebrate embryo. Bone replaces cartilage in older animals by a process of ossification. Different types of cartilage are recognized depending upon the relative amounts of the constituents of the chondrin. Hyaline (gristle) rings occur in the windpipe (trachea) and elastic cartilage is found in the ears and nose. Cartilage also occurs around joints and between the discs of the vertebrae where it has a protective function and can act as a shock absorber.

cartilaginous fishes *see* **Chondichthyes**.

casein a group of proteins containing phosphate, found in milk and cheese. It is precipitated as curd by the action of rennin (an ENZYME) and calcium.

catabolism the degradational processes of METABOLISM. For example, the breakdown of glucose during GLYCOLYSIS is a catabolic process.

catalyst a substance that increases the rate of a chemical reaction but can be recovered unchanged at the end of the reaction. All biological processes in animals and plants involve natural catalysts called ENZYMES.

catecholamine any one of a number of AMINES derived from tyrosine. Important ones such as dopamine, ADRENALINE and NORADRENALINE act as NEUROTRANS-MITTERS and HORMONES.

cathode the negatively charged electrode of an electrochemical cell to which CATIONS travel and gain electrons (i.e. where REDUCTION occurs).

cation a positively charged ion formed by an atom or group of atoms that has lost one or more electrons.

caudal of the tail, e.g. caudal vertebrae, caudal fin.

cell the basic unit of all living organisms. In most cases, cells are microscopic and may form unicellular organisms as in BACTERIA, or merge to create tissues or colonies. The two types of cell are PROCARYOTE and EUCARYOTE, the former being the more primitive.

cell division the process by which two daughter cells are formed from one parent cell with the nucleus dividing first (*see also* MITOSIS and MEIOSIS).

cell membrane (*also* **plasma membrane, plasmalemma**) the membrane that surrounds any cell, composed mainly of LIPID and PROTEIN and varying in thickness from 5–10 nm. It defines the outer limit of a particular cell. Some cells may have more than one membrane. Cell membranes have a structural role and also participate in the formation of the cell wall. The cell membrane is a major regulator of the transport of materials into and out of the cell.

cellulose a polysaccharide occurring widely as cell walls in plants. It is found in wood, cotton and other fibrous materials and comprises chains of GLUCOSE units. It is used in the manufacture of paper, plastics and explosives.

cell wall present in plant and bacterial cells, but not animal cells. In plants it consists of a matrix of POLYSACCHARIDE and PROTEIN with cellulose fibres embedded within it. In young plants, a thin primary cell wall is secreted first and then sometimes a stronger secondary cell wall is laid down on the inside of this once growth has finished. The secondary cell wall may be strengthened, as in woody plants (with LIGNIN) or the primary cell wall itself may have hardening substances secreted into it. The primary cell walls of adjacent cells are separated by a thin layer called the middle lamella composed of PECTINS. Channels through the walls connect adjacent cells and the plasma membrane (*see* CELL MEMBRANE) lies internal to the cell wall. In fungi the cell walls contain CHITIN, and in bacteria consist of AMINO ACIDS and polysaccharides in polymeric form. They provide support for these cells.

Celsius scale *or* **centigrade scale** (C) a temperature scale with a freezing point of 0° and a boiling point of 100°, devised by the Swedish astronomer Anders Celsius (1701-44).

Cenozoic the geological era from the end of the Mesozoic, 65 million years ago, to the present day, which is subdivided into two periods, the TERTIARY and QUATERNARY (*see* APPENDIX 5). Also called the *Age of Mammals* as it was during this time that these animals diversified and expanded. However, birds and flowering plants also became prominent.

central nervous system (CNS) the part of the nervous system that receives, co-ordinates and transmits all neural signals. In invertebrates such as

Polychaetes, it consists of a pair of anterior ganglia (brain) connected to longitudinal nerve cords which produce additional ganglia in each segment. In vertebrates the central nervous system consists of the brain and SPINAL CORD and information is received through the spinal nerves. Vertebrate spinal reflexes are local reflex arcs which enable rapid automatic responses to occur when particular stimuli are applied, without the involvement of higher centres in the brain.

centre of symmetry a concept of symmetry used in crystallography, the centre of symmetry is a central point around which all crystal features occur in equidistant and equal pairs, i.e. like faces, edges etc. are arranged in pairs in corresponding positions on opposite sides of the central point. The cube has an obvious centre of symmetry.

centrifugation a technique in which a high-speed rotating machine, a centrifuge, generates centrifugal forces to separate the various components of a liquid. Different components suspended in the liquid will separate at different centrifugal speeds depending upon their size and mass.

centriole a structure that is found in animal cells, but is absent from those of higher plants. They are very similar or identical to the basal bodies which occur at the bases of FLAGELLA or CILIA in those cells where these occur. Centrioles are hollow cylinders consisting of micro-tubules and normally occur in the cell in pairs at right angles to each other, near the nucleus. They have a role in cell division (*see* MITOSIS) and during this process move to either end of the nucleus to form the ends of the spindle, the fibres of which develop from them. Centrioles appear to be self-duplicating and contain their own DNA.

centromere the constricted region of a CHROMOSOME, to which a pair of sister CHROMATIDS are attached. All nuclear chromosomes of EUCARYOTIC cells have a centromere, the exact position of which varies but which is genetically inheritable. The centromere can therefore be used in the identification of chromosomes, which, to be observed under the microscope, must be in cells which are undergoing either MEIOSIS or MITOSIS. The centromere is also an important factor in nuclear division, as the absence of one will lead to failure of segregation; that is, one daughter cell will contain both sister chromosomes instead of one chromosome per each separate daughter cell. Thus only one cell is inheriting the genes that code for vital proteins.

Cephalopoda the most advanced class of the phylum Mollusca with a highly developed brain and CENTRAL NERVOUS SYTEM contained within a cartilaginous sheath. The class includes the extinct AMMONITES, nautilus, squids, cuttlefish and octopuses. The ammonites had well-developed external spiral shells, and the animal inhabited the last chamber, but *Nautilus* is the only living form to exhibit this structure. In the modern forms the shell is much reduced as in squids, internal as in cuttlefish or absent as in octopuses. Most cephalopods are marine and predatory with a clearly defined head surrounded by a ring of very mobile tentacles used to seize their prey. They swim by jet propulsion— water is expelled from the mantle cavity through a muscular siphon derived from the foot. Cephalopods have complicated and highly developed eyes and demonstrate cognitive powers that have been the subject of much research. Cuttlefish show remarkable colour changes, possibly as a means of communication.

cerebellum a part of the vertebrate HIND BRAIN which lies above the *medulla oblongata* and consists of two joined hemispheres. In mammals it is covered in a CORTEX, a much folded layer of GREY MATTER, and internally lies a core of WHITE MATTER. Muscle activity is controlled by the cerebellum, as is the maintenance of posture and balance.

cerebral cortex *or* **pallium** a much folded layer of GREY MATTER, the outer covering of the cerebral hemispheres of the CEREBRUM in vertebrates. It contains many SYNAPSES and the folding increases the surface area in which these can occur. It is at its mostly highly developed in mammals and processes and controls the neural information involved in the senses of touch, hearing and sight. It also controls the voluntary movement of muscles and contains centres associated with intellectual activity and memory.

cerebrospinal fluid a colourless, clear fluid with a composition similar to LYMPH, that bathes the surfaces of the spinal cord and brain and fills the cavities and ventricles in the CENTRAL NERVOUS SYSTEM. It contains some blood cells and supplies nutrients and has a protective function acting as a shock absorber to help prevent mechanical injury to the central nervous system. It is secreted by the choroid plexuses (projections from the epithelium) into the ventricles of the brain and reabsorbed by veins.

cerebrum the largest part of the vertebrate brain, consisting of the paired cerebral hemispheres covered by the CEREBRAL CORTEX, beneath which is white matter. It arises from the forebrain of the vertebrate embryo and controls complicated sensory and motor activities; learning, thought and memory. There may be as many as 10 billion nerve cells in the mammalian cerebral cortex, which is dominant both in terms of its size and function.

Cestoda a class of the phylum Platyhelminthes comprising the tapeworms which have a parasitic (*see* PARASITE) life cycle inside a host (usually vertebrate) gut. The head (scolex) bears suckers often with hooks by means of which it attaches to the intestinal lining of the host. Behind the head a series of "sacs" (proglottids) are produced containing male and female sex organs which give a segmented appearance. There is no gut and nutrients are absorbed from the host gut over the whole body surface which is protected by a tough, outer CUTICLE. The animals are HERMAPHRODITE and their life cycle requires both a primary and secondary host. In one tapeworm that affects man, *Taenia saginata*, the primary host, containing the mature worm, is man. Mature proglottids laden with fertilized eggs are passed to the outside in faeces and contaminate water or food (grass) subsequently eaten by cattle. In the intestine of the cow the egg develops into an embryo or larva which bores through the intestinal wall, entering the blood stream, from whence it is carried to muscle. Here it encapsulates and becomes a cyst. Another human being becomes infected by eating poorly cooked meat containing cysts from which the larvae emerge, mature, and repeat the life cycle.

Cetacea the order of marine mammals containing

the dolphins and whales. Characterized by fore-limbs developed as flippers, a powerful, flattened tail with two flukes adapted for swimming and a dorsal fin. A thick layer of blubber (fat), which serves as insulation and provides a food store, is present beneath a nearly hairless skin. Air is drawn into the lungs through a DORSAL blowhole which can be shut off when the animal dives. There are two sub-orders, Odontocetti, the toothed whales which are predators, e.g. dolphins and killer whales, and Mysticetti, PLANKTON feeders such as the blue whale which filter vast quantities of sea water through whalebone plates. The blue whale is the largest known living animal, in excess of 150 tonnes in weight and 30 metres (100 feet) in length.

CFC (*abbreviation for* chlorofluorocarbon) a chemical widely used in manufacturing processes, which reacts with and destroys OZONE, causing depletion of the OZONE LAYER.

chaeta (*plural* **chaetae**)a thin bristle made of CHITIN projecting from the skin of ANNELID worms and used in locomotion. In POLYCHAETES chaetae project from structures known as parapodia.

chalk a white, fine-grained, porous limestone formed from calcium carbonate and calcareous skeletal remains of micro-organisms. Chalk deposited in the Upper Cretaceous (*see* APPENDIX 5) covers much of northwest Europe.

character a particular inherited characteristic possessed by an organism, e.g. body colour in the fruit fly (*see also* DROSOPHILA).

chelation a reaction between a metal ION and an organic molecule that produces a closed ring, thus tying up the unwanted metal ion. Chelation occurs naturally in soils, removing metal ions in solution which may be potentially toxic to plants. This principle can be applied to domestic products, e.g. chelating agents are often added to shampoos to soften water by "locking up" calcium, iron and magnesium ions (*see* HARD WATER.)

Chelonia an order of the class REPTILIA that contains terrapins, turtles and tortoises. The body is encased in a bony shell, or covered by horny plates which are epidermal (*see* EPIDERMIS) in origin and the pelvic girdle and shoulder girdle are within the rib cage. Both marine and terrestrial forms occur. The jaws are developed as horny beaks without teeth and they are an ancient group, seemingly little changed since their evolution from stem reptiles in the Mesozoic era (*see* APPENDIX 5).

chemical bond *see* **bond**.

chemical equation the representation of a chemical reaction using SYMBOLS for atoms and molecules.

chemical oxygen demand (COD) an indicator of water (or effluent) quality. Oxygen demand is measured chemically (*compare* BIOLOGICAL OXYGEN DEMAND) using potassium dichromate as the oxidizing agent. The OXIDATION takes two hours, providing a much quicker assessment than the BOD.

chemisorption *see* **adsorption**.

chemistry the study of the composition of substances, their effects upon one another and the changes which they undergo. The three main branches of chemistry are ORGANIC, inorganic, and physical chemistry.

chemoreceptor a site within an organism that is sensitive to specific chemical stimuli, e.g. those involved in the senses of taste and smell. The carotid body, which occurs near a branch of the internal and external CAROTID ARTERY, contains chemoreceptors that are sensitive to pH levels and is involved in the control of reflex changes in the rate of RESPIRATION.

chemotaxis the movement of an organism towards or away from a specific chemical. For example, some bacteria in a solution have been shown to move (using their flagella) towards an area of high glucose concentration.

chemotherapy the use of toxic chemical substances to treat diseases where the chemicals are directed against the invading organisms or the abnormal tissue.

chemotropism the movement or growth, generally of part of a plant or plant organ, in a direction determined by the concentration gradient of a particular chemical that is present. An example is the growth of a pollen tube downwards into the stigma, in response to the presence of sugars, during fertilization.

chimaera an individual with tissues of two or more different GENOTYPES. The cause may be mutation in a developing embryo. In plants it may be because of mutation or GRAFTing.

chirality in STEREOISOMERISM the idea of left and right handedness applied to chemical molecules (*see also* OPTICAL ACTIVITY). The structure of a molecule is chiral if it cannot be superimposed on its own mirror image.

Chiroptera an order of mammals comprising the bats, the main characteristic of which is flight. This is effected by means of membranous wings spread between the elongated forelimbs and fingers, and the hind limbs and, occasionally, the tail. Bats are usually nocturnal and insectivorous, with large ears adapted for ECHO LOCATION, by means of which they locate prey and sense their surroundings. Some are nectar feeders (and plant pollinators), others eat fruit, and vampire bats suck blood.

chi-squared test a statistical test used to determine how well data obtained from an experiment (the observed data) fits with the data expected to occur by chance. The chi-squared test is a simple method of checking that the experimental results are significant and have not just arisen from chance events.

chitin a HYDROCARBON, related to CELLULOSE but containing nitrogen, that forms the skeleton in many invertebrates, e.g. the shells of insects.

chlorofluorocarbon *see* **CFC, ozone layer**

chlorophyll the green pigments of plants present in the cell ORGANELLES called CHLOROPLASTS. Chlorophyll is an essential factor in PHOTOSYNTHESIS as it traps energy from sunlight and uses it to split water molecules into hydrogen and water.

Chlorophyta a division of ALGAE, the green algae, the largest and most varied group. The cell wall is of cellulose and they contain CHLOROPHYLLS a and b. CHLOROPLASTS (organelles that carry out photosynthesis) are present and store starch as a food reserve. Single-celled forms occur, sometimes with FLAGELLA for movement, or they may exist in colonies. Filamentous (e.g. *Spirogyra*) multicellular types exist and also forms with a flattened multicellular *thallus*. They are mainly aquatic, occurring in both marine and fresh water situations, but can flourish in a terrestrial environment if the surroundings provide sufficient water.

chloroplast an ORGANELLE that is found within the cells of green plants and ALGAe and that is the site of PHOTOSYNTHESIS. The chloroplast consists of closed membrane discs called thylakoid vesicles (often stacked together to form a granum), which are surrounded by a watery matrix termed the stroma. The stroma contains enzymes necessary for the CALVIN CYCLE, and the thylakoid vesicles contain chlorophyll, an essential pigment of photosynthesis as it absorbs light energy.

cholesterol an insoluble molecule that is abundant in the plasma membrane of animal cells. In mammals, cholesterol is synthesized from saturated fatty acids in the liver and is transported by carrier molecules in the blood, called low-density LIPOPROTEIN (LDL). Cholesterol enters cells via the LDL-receptor on the plasma membrane of the cell. This mechanism of uptake helps regulate the levels of cholesterol in the blood. If a person's diet is high in cholesterol, the number of LDL-receptors on the cell membrane will decrease, which will result in a decrease in the cellular uptake of cholesterol, leading to a corresponding increase in the levels of cholesterol in the blood. This excess cholesterol is deposited in the artery walls and ultimately leads to ARTERIOSCLEROSIS. Some unfortunate individuals are particularly prone to arteriosclerosis as they carry a gene that codes for a defective LDL-receptor, making the cellular uptake of cholesterol an impossible task. This inherited disorder is called familial hypercholesterolemia, which basically translates as "too much cholesterol in the blood." Cholesterol is also the main component of steroid hormones and bile salts.

choline a base ($CH_2OHCH_2N(CH_3)_3OH$) that occurs in certain types of PHOSPHOLIPIDS. It is found in the brain, egg yolk, and bile in this form. In the form of ACETYLCHOLINE it is an important substance in the transmission of nerve impulses (*see also* SYNAPSE).

Chondrichthyes a class of vertebrates comprising the cartilaginous fishes, once thought to be a more primitive group than the BONY FISH but now recognized to be highly specialized. They are mostly marine and have a skeleton composed entirely of cartilage. The fishes are covered by tiny placoid scales, each extended into a point and composed of ENAMEL on the outside and dentine on the inside, like the mammalian tooth. A pair of nostrils is present but these open into blind-ended cups and are not used for respiration but for olfaction. A LATERAL LINE system is present but there is no SWIM BLADDER. The male fish possess a pair of copulatory organs, the claspers, developed from the pelvic FINS. A common chamber, the cloaca, with a single external opening, is formed from the end of the hind gut into which lead the excretory and reproductive systems. A salt-secreting rectal gland is attached to the cloaca. Parts of the NOTOCHORD (an embryonic feature) persist in the adult but only between adjacent vertebrae. Two sub-classes are recognized—the Holocephali (Chimeras), e.g. ratfish, which possess an OPERCULUM, but not SPIRACLES. The rest of the cartilaginous fish, e.g. rays, sharks, skates, belong to the sub-class Elasmobranchii and have no operculum but possess conspicuous spiracles.

Chordata a phylum of animals that possess a NOTOCHORD at some stage in their development and have a hollow dorsal nerve cord. GILL SLITS opening from the pharynx are likewise present at some developmental stage, and a post-anal tail. The subphylum vertebrata is the most important group.

chorion in reptiles, birds and mammals it is one of three extra embryonic membranes surrounding the ALLANTOIS and embryo yolk sac (*see* AMNION). In insects it is secreted by the OVARY and forms the protective shell surrounding the egg.

chromatid one of a pair of side-by-side replica CHROMOSOMES joined by the CENTROMERE and produced during the replication of DNA within a cell.

chromatography a method for isolating the constituents of a solution by exploiting the different bonding properties of molecular substances. One complex method is affinity chromatography, which is used to separate proteins. A specific ANTIBODY is coupled to small plastic beads to form a column through which the protein solution is passed. The only protein that will bind to the column is the one recognized by the antibody; other proteins will pass through the column unimpeded.

chromatophore a cell containing pigment, usually found in the outer covering of some CRUSTACEA or in the skin of lower vertebrates (e.g. chameleon). The animal can effect colour changes by dispersal or concentration of pigment in the chromatophores. A common type of chromatophore in vertebrates is the one containing MELANIN.

chromoplast one of a number of pigment-containing organelles present in plant cells.

chromosome a highly structured complex of DNA and PROTEINS found within the nucleus of EUCARYOTES. A nuclear chromosome is involved in the following three functions:

(1) Replication—this ensures the correct duplication of the DNA, or cells would become unviable through genetic mutation.

(2) Segregation—this ensures that the newly replicated chromosomes will separate and that each will become part of a daughter cell. It is a very important function dependent largely on the CENTROMERE.

(3) Expression—this ensures that the genes present on the chromosome are correctly transcribed to preserve the genetic information coding for specific proteins.

PROCARYOTIC cells contain a single, circular, chromosomal DNA molecule, whereas evidence indicates that EUCARYOTIC cells contain a single, linear molecule of double-helical DNA within their nucleus. However, the eucaryotic ORGANELLES, MITOCHONDRIA and CHLOROPLASTS contain a single circular molecule of DNA, which perhaps is an indication of their origin, i.e. as procaryotic cells that have evolved into eucaryotic organelles. A species will have a characteristic number of base pairs of DNA all packed into a characteristic number of chromosomes. For instance, yeast cells contain approximately 14×10^6 base pairs (bp) packed into 16 chromosomes, and the human nucleus contains approximately 3000×10^6 bp packed into 46 chromosomes. Chromosomes are visible under the microscope only during MEIOSIS or MITOSIS, when they condense to form short thick structures before separating.

chyme the semi-solid mass of partly digested food that leaves the vertebrate stomach to enter the small intestine.

cilia (*singular* **cilium**) fine thread-like structures on cell surfaces, which beat to create currents of liquid over the cell surface or to move the cell.

circadian (rhythm) regular changes occurring in the activity, behaviour or physiology of animals or plants approximately every 24 hours. The sequence may continue for days even in the absence of normal, repetitive events, e.g. daylight and darkness, with which it is linked.

citric acid $C_3H_5O(COOH)_3$ a tricarboxylic acid that occurs in many fruits, especially lemons. It is now produced commercially by fermentation of sugar by moulds of the *Aspergillus* genus and is itself used as a food flavouring and in the production of effervescent salts and drinks.

citric acid cycle a complex set of biochemical reactions controlled by ENZYMES. The reactions occur within living cells, producing energy, and the cycle is instrumental in the final stages of the OXIDATION of CARBOHYDRATES and fats and is also involved in the synthesis of some AMINO ACIDS.

cladistics a method of TAXONOMY in which organisms are classified into groups according to the order in time when they arose from a common ancestral line. It assumes that new species arise suddenly from a common ancestor, rather than by gradual evolutionary changes (*compare* NEO-DARWINISM).

class a category used in the CLASSIFICATION of organisms, e.g. class Mammalia.

classification any method of grouping organisms together to reflect their relationships to one other, based on a number of different criteria. Various methods of classification have been employed and studied, and in all a hierarchy emerges. The higher the position of a category in the hierarchy, the fewer the similarities between its member organisms. In animals the ascending order of classification is species, genus, family, order, class, phylum, kingdom (*see individual entries*). In plants, division replaces phylum and species belonging to the same genus share the greatest number of similar characteristics.

clavicle a bone of the pectoral (shoulder) girdle which is the collarbone of man.

cleavage describes the stage in early embryonic development when the fertilized egg divides several times to become a multicellular embryo (*see* BLASTULA).

climatic zones regions of the Earth possessing a distinct pattern of climate, each region approximating to belts of latitude. There are essentially eight zones. Working from the two polar caps of snow there is a zone of boreal (northern) climate in the northern hemisphere with a range of temperatures, two belts of humid temperate conditions, two subtropical arid/semi arid zones and finally the humid tropical climate near the equator.

clone an organism, micro-organism or cell derived from one individual by asexual process. All such individuals have the same GENOTYPE. Plants propagated by cuttings are clones.

closed-chain in organic chemistry when carbon atoms are bonded together to form a ring, as typified by BENZENE. Closed-chain compounds are thus called ring or cyclic compounds (*see also* HETEROCYCLIC COMPOUNDS).

clotting factors *or* **coagulation factors** a number of substances found in blood plasma that in particular circumstances become involved in a chain of chemical reactions leading to the formation of blood clots. They are commonly identified by Roman numerals, e.g. factor VIII, although each has a specific name. Lack of any of the clotting factors, perhaps due to a genetic disorder, e.g. haemophilia, results in the loss or reduction of the blood's ability to clot.

cloud water droplets or ice formed by condensation of moisture in rising warm air. Rising on convection currents, the warm moist air reaches the higher, cooler conditions and condenses. Clouds are classified on their form into three major groups—cumulus (heap), stratus (sheet) and cirrus (fibrous) which are further sub-divided into genera and species on features such as transparency, type of growth and arrangement.

Cnidaria *or* **Coelenterata** a phylum of mainly marine, aquatic invertebrates that show RADIAL SYMMETRY. They are diploblastic, i.e. the body wall consists of two layers of cells separated by a jelly-like layer, the mesoglea, which in simple forms is non-cellular but may contain cells and connective tissue fibres. Cnidarians have a central digestive sac, the gastrovascular cavity with a single opening serving both as mouth and anus. The opening is surrounded by a ring of tentacles bearing stinging cells characteristic of the group, the *cnidocytes*. The animals are carnivorous, employing the toxin-secreting cnidocytes to immobilize prey, and the tentacles to push it into the central digestive opening. Cnidarians show two forms, POLYPs and MEDUSAe, both based on the body plan outlined above. One species may exhibit both forms through the life cycle showing an alternation of generations. Polyps are cylindrical and sessile; they may be colonial as in corals or solitary as in sea anemones. Medusae are bell-shaped and free swimming and may contain gonads; they are typical of the jellyfish. Reproduction in the Cnidaria is both ASEXUAL (budding) and sexual via gametes (fragmentation). There are three classes, Hydrozoa, Scyphozoa (jellyfish) and Anthozoa (sea anemones, corals).

cnidocyte a cell, unique to the phylum CNIDARIA, that is used in defence and capturing prey. Every cnidocyte has a stinging capsule, the NEMATOCYST, which contains a thread that can be shot out to wrap around prey.

CNS *abbreviation for* CENTRAL NERVOUS SYSTEM.

coagulation factors *see* **clotting factors**.

coal *see* **fossil fuels**.

coccus (*plural* **cocci**) a spherical BACTERIUM which may occur singly or grouped together in chains, e.g. *Streptococcus*, or as a mass, e.g. *Staphylococcus*. Both these types are important to humans since they contain PATHOGENIC species.

coccyx the terminal region in the vertebral column (tail) of apes and humans, formed from the fusion of 3–5 caudal vertebrae.

cochlea a part of the INNER EAR in some reptiles, birds and mammals, comprising a complex organ, spirally coiled like a snail shell. It contains three canals, an upper vestibular canal and a lower tympanic canal both filled with fluid called perilymph. In between lies a smaller canal, the cochlear duct, also fluid-filled with endolymph. The whole is contained within the labyrinthine cavity of the temporal bone. The organ of corti within the cochlear duct contains highly specialized cells bearing hairs which receive and transform sound waves into nerve impulses which are then carried to the brain. Sound waves cause pressure changes within the fluid and membranes of the cochlea that trigger the activity of the hair cells. The base of the cochlea is more sensitive to higher frequency sound waves and the top responds to lower frequencies.

COD *abbreviation for* CHEMICAL OXYGEN DEMAND.

coelacanth a type of bony fish belonging to the order Coelacanthiformes, sub-order Crossypterygii, the flesh-finned fishes. All were thought to be extinct, but a living representative was discovered in 1938. Living coelacanths are members of the genus *Latimeria* and characteristically have a three-lobed tail fin. They are large fish living at great depths on the floor of the Indian Ocean and have crutch-like pectoral fins used for "walking." Coelacanths give birth to live young and do not possess lungs. They are a remnant of numerous kinds of lobe-finned fishes that were present in the DEVONIAN period. However, these extinct forms were freshwater animals with lungs and some were the ancestors of the amphibians.

Coelenterata *see* **Cnidaria**

coelom the main body cavity of most invertebrates and all vertebrates, which is fluid-filled and contains the gut and other internal organs. The coelom is formed from embryonic mesoderm lined by tissue derived from this. Specialized ducts, sometimes bearing CILIA (coelomoducts), lead from the coelom to the exterior allowing for the elimination of waste material and gametes. In higher animals such as vertebrates these are further specialized. In vertebrates, the fluid in the coelom helps to protect organs from injury, and in invertebrates such as

ANNELIDS it provides a hydrostatic skeleton.

cofactor an inorganic substance or non-protein organic substance that must be present before a particular ENZYME reaction can occur. If the cofactor consists of organic molecules it is known as a coenzyme. Cofactors may take part in the enzyme reaction or operate by activating the enzyme itself.

cold-bloodedness *see* **poikilothermy**.

Coleoptera a vast order of the class INSECTA, comprising the beetles and weevils of which there are at least 500,000 known species, the largest animal order. They exploit a large number of different habitats, some feeding on vegetation and decaying organic material, while others are predatory and carnivorous. The paired forewings are horny, protective sheaths (elytra) that meet precisely along the mid-dorsal line. These protect the abdomen and delicate, membranous hindwings, which are sometimes used for flying but are reduced or absent in other species. The mouth parts are often enlarged for biting, and both young (larvae) and adult beetles of certain species can be serious economic pests of crops, timber and stored food. Many are beautifully coloured and patterned, e.g. the ladybird beetles.

collagen an important fibrous PROTEIN that forms almost one third of total body protein in mammals. It is found in tendons, bone, skin and cartilage and comprises mainly glycine, proline (AMINO-ACIDS) and hydroxy-proline in helical structures.

colligative property an aspect of a solution that depends solely on the concentration and not the nature of the dissolved particles. Colligative properties are important when determining, say, osmotic pressure (*see* OSMOSIS), vapour pressure, or the freezing point of a solution.

colloid a substance forming particles in a solution varying in size from a true solution to a coarse suspension. The particles, measuring 10^{-4} to 10^{-6} mm, are charged and can be subjected to ELECTROPHORESIS. Many substances such as vegetable fibres, rubber and proteins occur naturally or are at their most stable in the colloidal state. Different types of colloids are recognized—gels, emulsions, sols, AEROSOLS and foams.

colon the part of the vertebrate large intestine that lies between the RECTUM and the CAECUM. In mammals, water is absorbed from food residues in the colon and bacteria contained within it produce VITAMINS, especially VITAMIN K.

colony a group of individuals of the same species living together and to some extent interdependent. In some cases the individuals are actually joined, as in corals, and function as one. In others, e.g. insects, they are separate but exhibit a high degree of organization, possibly with specialization of functions. Also a colony may be a BACTERIUM or yeast growing from a parent cell on a food source.

colour blindness the inability to distinguish between certain colours, the most common being Daltonism where red and green are confused. This is an inherited disfunction, the recessive gene for which is carried on the X-chromosome, hence SEX LINKED and more likely to be present in men.

commensalism where individuals from different species live together such that one benefits from the association while the other is not affected.

compound eye the type of eye present in CRUSTA-

CEANs and insects, made up of numerous facets, light detectors called ommatidia. Each of these produces an image of a tiny part of the visual field so that the whole "picture" resembles a mosaic. The resultant image is less sharp than that produced by a mammalian EYE; however the compound eye is more effective at detecting movement as it can detect light flashes pulsing at a higher rate. Some insects, such as dragonflies, have been shown to have excellent sight rivalling that of vertebrates. Insects also have colour vision and in some (bees), sight extends into the ultraviolet range of the spectrum, which cannot be detected by human beings.

concentration the quantity of substance (SOLUTE) dissolved in a fixed amount of SOLVENT to form a SOLUTION. Concentration is measured in moles per litre (mol l^{-1}).

condyle the round, smooth knob of bone that fits into the socket of an adjacent bone, the whole forming a JOINT. The joint allows for side-to-side movement and up-and-down but not for rotation. Condyles on either side of the head attaching the lower jawbone to the SKULL allow for chewing movements. The occipital condyles articulate the skull to the vertebral column and allow for movement of the head.

cones these are light-sensitive RECEPTOR cells found in the vertebrate RETINA which are involved in the detection of colour. They require bright light in order to function and hence tend to be absent from the eyes of nocturnal animals. Cones contain the pigment retinene and the protein opsin, and there are three types that, in bright light, bleach at different wavelengths (blue, green, red) enabling colours to be detected (*see also* EYE, ROD CELL).

congenital a term used to describe a property present at birth, which may be an inherited genetic condition such as DOWN'S SYNDROME, or be caused by environmental factors such as drugs (e.g. thalidomide). Bacteria and viruses can also cause congenital infections or defects e.g. the virus *Cytomegalovirus*. The causes of some congenital abnormalities may not necessarily be known as in the case of some heart defects.

conjugation a form of sexual reproduction found in simple organisms where two cells come together to exchange genetic material. It is found in some bacteria, e.g. *Escherichia coli*, some ciliated PROTOZOANS, e.g. *Paramecium*, and some algae such as *Spirogyra*. Conjugation is sometimes achieved by means of a tube, the conjugation tube, which grows out from one or both of the participants.

connective tissue a common tissue found in animals that is further differentiated into various types depending upon the proportions of its constituent materials and its function. It is derived from embryonic MESENCHYME and consists of a lot of extracellular non-living matrix in which is spread a sparse number of cells, such as MAST CELLS and fibroblasts, and fibres, such as COLLAGEN, elastic and reticular fibres. Connective tissue frequently has a good blood supply and is often bathed in tissue fluid. Different types found in vertebrates include loose connective tissue (packing material holding organs in place and to bind tissues together), ADIPOSE TISSUE, CARTILAGE, BONE, BLOOD and LYMPH. Also, fibrous connective tissue occurs in TENDONS and LIGAMENTS.

continental drift a geological concept formulated by the German geophysicist Alfred Wegener (1880-

1930) that 200 million years ago the earth consisted of a large single continent, called PANGAEA, which broke apart to form the present continents. An explanation of how such huge land masses move is provided by studying the vast plates that make up the outer layer of the earth, called the crust. The crustal plates are believed to float on a partially molten region of the earth between the crust and the earth's core, the lower mantle (hence the modern term plate tectonics).

continental shelf the surface between the shoreline and the top of the continental slope where the gradient steepens at a depth of approximately 150 metres (500 feet). The average width of the shelf is 70 metres (230 feet).

contractile vacuole a cavity in a cell, surrounded by a membrane, which takes in water and expands, and then suddenly contracts and expels the fluid to the outside. It is typical of freshwater PROTOZOA, some freshwater sponges and some flagellated algae. Contractile vacuoles are usually found close to the cell membrane, and drainage channels from the cytoplasm appear to lead into them so that they fill. At some point the membrane of the vacuole seems to join with the cell membrane and a pore forms through which the water is rapidly expelled. This process has an obvious role in OSMOREGULATION (and excretion).

Copepoda a large sub-class of the class CRUSTACEA with both marine and freshwater forms. They are all small (0.5–2 mm long) and are without a carapace or COMPOUND EYES. The thorax bears five or six pairs of APPENDAGES used for swimming but none occur on the abdomen. Copepods are important species in the PLANKTON and may be free living filter feeders while some are parasitic (e.g. fish louse). *Cyclops* is a well-known freshwater genus with species having one median eye (OCELLUS).

coral *see* **Actinozoa**

coral reef a hard bank built up of the carbonate skeletons of colonial corals (and algae). There are several forms of reef including BARRIER REEFS, fringing reefs (which are attached to the coast) and atolls (where a reef encloses a lagoon). Certain conditions are essential for the growth of reefs including temperature, a maximum water depth of 10m, clear water with no land-derived sediment and normal marine salinity.

corm an organ of vegetative reproduction, present in some plants such as *Gladiolus* and *Crocus*, formed underground from a swollen, short stem. This acts as a food store and is surrounded by protective scale leaves which are the remains of the previous season's growth. One or two buds formed in the axils of the scale leaves grow up in the spring to form new foliage and flowers, using the stored food reserve, which shrinks. As the season advances, the foliage leaves wither and the stem at the base swells to form a new corm (*see* BULB).

cornea the outer, transparent, exposed layer of the vertebrate and cephalopod (squid) EYE, over the IRIS and LENS, and continuous with the sclera which surrounds the whole eyeball. It is a connective tissue layer containing COLLAGEN and without a blood supply. The cornea is supplied with nutrients from the AQUEOUS HUMOUR, a fluid that occupies the space between itself and the iris and lens. The cornea refracts the light entering the eye, acting as a coarse focus directing rays to the lens.

coronary artery one of the two main vessels that carry oxygenated blood to the heart muscle. In general terms, the left coronary ARTERY divides into two branches and supplies most of the blood required by the left VENTRICLE, while the right coronary artery supplies blood to the right ventricle. The two coronary arteries originate from the AORTA, but run independently on to the surface of the heart.

corpus luteum this is formed in the cavity created in a GRAAFIAN FOLLICLE after the release of an egg (OVUM) from the OVARY. The follicle tissue grows to form a solid, yellow mass which is responsible for the secretion of the HORMONE, PROGESTERONE. If the egg is not fertilized and no pregnancy ensues, the corpus luteum disintegrates and another follicle matures and releases an egg. However, it persists in the pregnant mammal under the influence of chorionic gonadotrophin, a hormone secreted by the placenta (*see* OESTRUS CYCLE).

cortex in animals this is the outer tissue layer of various organs, e.g. the CEREBRAL CORTEX, adrenal cortex of the ADRENAL GLANDS and renal cortex of the KIDNEYS. In plants it describes the layer in roots and stems that lies betwen the VASCULAR SYSTEM and the EPIDERMIS. It is composed of PARENCHYMA cells derived from the region at the tips of shoots and roots where rapid cell divisions occur, the apical meristem. The term also applies to a tissue layer present in some algae and lichens.

corticosteroid any one of a number of STEROID HORMONES produced in the CORTEX of the ADRENAL GLAND. There are two types—mineralocorticoids, such as ALDOSTERONE, which regulate water and salt balance, and GLUCOCORTICOIDS (e.g. CORTISOL, cortisone) which act primarily on glucose metabolism.

cortisol *or* **hydrocortisone** a principal GLUCOCORTICOID in many mammals, responsible for glucose metabolism. It is one of the HORMONES released in response to stress and causes a rise in blood pressure. It is involved in the regulation of fat deposition in the body and is medically important, being used to treat certain skin conditions, allergies and rheumatic fever. Cortisone is a very similar hormone in all respects, and is also the name given to synthetically produced cortisol.

covalent bond the joining of two atoms by means of the equal sharing of their electrons. In a single covalent bond, one electron pair is equally shared between two atoms (as in the hydrogen molecule, H_2). In a double bond, two electron pairs are shared, and in a triple bond, three electron pairs are shared between two atoms. Covalent bonds are usually formed between non-metallic elements in which the atoms have a strong but equal attraction for electrons, resulting in the formation of strong, stable bonds.

cranial nerves in vertebrates these are 10 to 12 pairs of nerves arising directly from the BRAIN, each with dorsal and ventral branches (roots) that remain separate. Each root is assigned a separate roman numeral and also a name and some cranial nerves are mainly sensory while others are largely motor. Examples are I and II, which are mainly sensory—the olfactory and optic nerves, and XII, a motor nerve (the hypoglossal) serving the tongue. The cranial nerves and SPINAL NERVES are important elements of the PERIPHERAL NERVOUS SYSTEM.

cranium *or* **neurocranium** the part of the vertebrate skull that encloses and protects the BRAIN and
INNER EAR. It is formed from several flattened fused
bones that are joined by immovable suture joints.

Cretaceous the final geological period marking the
end of the Mesozoic era and extending from about
135 million years ago to 65 million years ago. Great
flooding occurred and large deposits of chalk were
laid down as the characteristic rock type in western
Europe. The flowering plants made their appearance and became abundant and in the early part of
the Cretaceous the DINOSAURS reached their peak.
However, by the end of the period, for reasons that
not fully understood, the dinosaurs and flying reptiles became extinct.

Crick, Francis H. *see* **Watson, James Dewey**.

Cromagnon man the earliest known modern man,
Homo sapiens, who is believed to have existed in
Europe about 35,000 years ago. They made and
used tools of bone and stone, and bequeathed contemporary human beings an intriguing record of
their presence in the haunting cave paintings at
Lascaux, in the Dordogne region of France. Fossils
of this hominid were first discovered at Cromagnon
in France in 1868.

crop an expanded part of the ALIMENTARY CANAL near
the anterior end, an enlargement of the oesophagus,
in which food is stored and may be partially digested. It is particularly associated with the birds,
in which it is a thin-walled sac that may produce
secretions, e.g. crop milk, in pigeons.

crossing-over the reciprocal exchange of genetic
material between two HOMOLOGOUS CHROMOSOMES
during MEIOSIS. This process occurs at a site called

the chiasma of the two chromosomes and is responsible for GENETIC RECOMBINATION.

crude oil *see* **fossil fuels, petroleum**.

Crustacea a class of ARTHROPODS containing many species, mostly living in marine and freshwater environments. Within the crustacea are the DECAPODA (lobsters, crayfish, crabs and shrimps); Isopoda (a small group including the "pill bugs"); the most numerous COPEPODA (planktonic forms); OSTRACODA (small filter feeders); Cirrepedia (the barnacles); MALACOSTROCA (prawns and shrimps), and the Branchiopoda (a primitive group including the water fleas). Crustacea usually have a well-defined head with an anterior pair of antennules bearing sense organs behind which lie a pair of ANTENNAE which may be modified for swimming or attachment. Mouthparts and COMPOUND EYES are also generally present. The body consists of a segmented thorax and abdomen, but these may not be well defined, and the outer layer may be a CHITINOUS CUTICLE or in the form of a carapace (shell). Typically, the body segments bear paired APPENDAGES known as biramous appendages. These are modified to perform a variety of different functions—swimming, feeding (sometimes filter-feeding), and as GILLS. The sexes are generally separate, and usually the eggs hatch to produce free-swimming nauplius larvae.

cuticle secreted by the EPIDERMIS and present in many plants and invertebrates, it is a non-cellular layer that has a strengthening and protective function. In plant stems the cuticle is a waxy outer layer which is formed of cutin, a mixture of fatty substances, which prevents the loss of water. In invertebrates, the cuticle may be composed of several layers and usually consists of CHITIN or COLLAGEN-like PROTEINS. The cuticle may remain soft but usually it is hardened to form a protective layer and to prevent water loss. In many arthropods it forms a hard EXOSKELETON, and in order to grow the animal has to moult by splitting the cuticle and extracting itself from its "shell" (ECDYSIS).

Cyanobacteria these were formerly classified as blue-green algae, which remains their common name, but they are now regarded as BACTERIA. They are PROCARYOTIC organisms and may be single-celled, occur in colonies or filaments. They are found in all aquatic environments and are often specialized, growing, for example, in hot springs. They possess the ability to PHOTOSYNTHESIZE and achieve this by means of chlorophyll a, CAROTENOIDS and phycobilins (proteins), dispersed throughout the cell, which also confer colour. Some species are able to fix nitrogen and hence are important in soil fertility. Reproduction is usually ASEXUAL but a form of sexual CONJUGATION may occur. Some cyanobacteria occur in the plankton and these may possess gas vacuoles for buoyancy. Others form symbiotic (*see* SYMBIOSIS) associations with other organisms, e.g. LICHENS. Blooms of some species in freshwater lakes enriched with agricultural fertilizers can be a problem as toxins are produced that kill fish and other animals that may drink the water.

cyclic AMP cyclic adenosine monophosphate, which is derived from ADP and ATP. It is widely found in animal cells and is a very important intermediary agent in numerous biochemical reactions that are controlled by HORMONES. When a particular hormone reaches its target cells, cyclic AMP production is activated by the ENZYME adenylate cyclase and its level rises. The cyclic AMP then goes on to activate further enzyme reactions which are under the overall control of the hormone concerned. Cyclic AMP additionally has a role in nerve transmission, cell division, immune responses and gene expression.

Cyclostomata *see* **Agnatha**

cytochrome one of a number of PROTEINS involved in electron-transfer (*see* ELECTRON TRANSPORT CHAIN) in the MITOCHONDRIA and CHLOROPLASTS of cells. A cytochrome has a haem group (iron-containing) and reversible changes in the iron atom between the reduced Fe (II) and oxidized Fe (III) states enable electrons to be transferred.

cytogenesis *or* **cytogeny** the formation and development of cells. **Cytogenetics** is the associated study involving cell structure and the structure and behaviour of chromosomes in relation to inheritance.

cytology the study of cells, their structure and function. Microscopy, particularly ELECTRON MICROSCOPY, has permitted the study of cell components, including the NUCLEUS and CHROMOSOMES.

cytoplasm all the contents of a cell, except the NUCLEUS or nuclei, including the plasma membrane (*see* CELL MEMBRANE). It consists of a jelly-like matrix in which various ORGANELLES and granules are dispersed. In some cells (e.g. *Amoeba*) the cytoplasm is arranged as a denser, outer plasmogel (ectoplasm) and an inner, more fluid plasmasol (endoplasm).

cytoplasmic inheritance a feature of some lower animals and plants whereby the inheritance of characters is transmitted by self-replicating ORGANELLES outside the nucleus.

cytosine

cytosine ($C_4H_5N_3O$) a nitrogenous base component of the NUCLEIC ACIDS, DNA and RNA, which has a pyrimidine structure and BASE PAIRS with GUANINE.

cytotoxic a term meaning lethal or poisonous to living cells. It usually refers to the T-CELLS (lymphocytes), which are a part of the vertebrate immune system, and also to drugs that are used medically to destroy cancer cells.

D

dark reaction the stage in PHOTOSYNTHESIS that occurs in the stroma of CHLOROPLASTS in plants and is not directly dependent on light (*see* CALVIN CYCLE).

Darwinism the theory of evolution described by the British naturalist Charles Robert Darwin (1809-1882). He first established the evolutionary theory of environmental forces acting as agents of natural selection on successive generations of organisms. The theory had five elements:

(1) Individuals have random variability, more so if they sexually reproduce.

(2) Reproductive capacity inevitably leads to competition both within (intra) and between (inter) species as populations tend to remain a set size (if there are no major external influences to upset this, e.g. mass-slaughter of animals by humans or an environmental catastrophe such as a huge oil slick).

(3) Some individuals are better adapted than others to their environment, which will help in their survival and reproductive success, i.e. fitness.

(4) Some of the characteristics that make parents successful in their ability to survive and reproduce are inherited by their offspring, thus increasing the probability of success for those offspring. These

characteristics become more widespread in the population. This is evolutionary change; the surviving species has adapted better to an environmental niche, i.e. survival of the fittest.

(5) The descendants of a single stock tend to diverge and become adapted to many different environmental NICHES. Darwin's paper, "Origin of Species" (1859), was greatly criticized by both the religious and scientific establishments of that period. He did not know of GENES as the units of inheritance, so it was understandable that he believed in the inheritance of acquired characteristics, as this would be a rational explanation for the phenomena he observed. With the knowledge that information regarding an organism's PHENOTYPE is present in their genes, then Darwin's theory can be summarized as follows—if the organisms have the capacity for genetic variability, then new species can emerge that will adapt to new environments, while the old species, which are no longer suited to the surrounding environment, will eventually die out.

DDT *abbreviation for* dichlorodiphenyltrichloroethane, $(ClC_6H_4)_2CH(CCl_3)$, a powerful pesticide much used in agriculture around the middle of the 20th century. It is a stable compound, which accumulates in the soil and becomes concentrated up the food chain. It is soluble in lipids and therefore builds up in the fatty tissues of carnivores, e.g. fish-eating birds. This progressive concentration through each stage of the food chain is called *biological magnification*. DDT has even been found in human breast milk. Its use is now restricted or, in some countries, banned.

Decapoda an order of predominantly marine crustaceans that are distributed worldwide. Both swimming and crawling varieties are found; prawns and shrimps the former, lobster and crabs the latter. The EXOSKELETON is hardened with calcium carbonate to form a carapace fused over the upper part of the animal, covering the thorax and head. A characteristic of the order, and the origin of the name, is five pairs of walking legs, the first of which are modified in the crawling varieties to form powerful pincers. The remaining three pairs of APPENDAGES are used for feeding or swimming. The females usually carry fertilized eggs until hatching.

decay the process whereby a radioactive (*see* RADIO-ACTIVITY) substance is transformed into its "daughter" products, and ultimately may form a stable atom.

deciduous the term applied to plants that shed their leaves at the end of the growing season. This happens in the autumn in temperate regions, and during the long dry season of the tropics. This is to prevent excessive water loss through transpiration (which in the dry tropics could exceed available supply).

decimal a structured system of numbers based on 10. *See also* APPENDIX 4.

decomposers any organism that breaks down dead organic matter, whether organisms or plant and animal waste, to obtain energy leaving simple organic or inorganic material. Earthworms, but mainly bacteria and fungi, fulfil this role. Carbon dioxide is released and the accompanying heat is useful, where conditions permit, in killing parasitic and pathogenic bacteria, eggs, etc (*see also* CARBON CYCLE and NITROGEN CYCLE).

decomposition the breakdown of a substance into simpler molecules or atoms by various means, e.g. pyrolysis (heat), ELECTROLYSIS, or biological agents (biodegradation).

degeneration the process, during evolution, whereby an organ undergoes a reduction in size becoming vestigial or disappearing completely. The human appendix is a case in point and now performs no function. Degeneration in some cases renders animals *apparently* more primitive than they really are. Degeneration of cells, tissues or organs may be because of disease and can result ultimately in the death of the affected part.

degradation the breakdown and simplification of complex molecules.

dehydration the removal of water. When preparing specimens for study under the microscope (*see* MICROSCOPY), it is necessary to eliminate water before the tissue is stained. It is achieved by immersion of the sample for several hours in successively stronger ethyl alcohols. This ensures that when the sample is embedded in paraffin wax it does not deteriorate. In chemistry, dehydration is the removal of water that is chemically held in a molecule or compound by the action of heat, often with a CATALYST or chemical acting as a dehydrating agent. Sulphuric acid is an example of the latter. In medicine, dehydration is the excessive, often dangerous, loss of water from the body tissues.

dehydrogenase a class of enzyme that facilitates the removal of hydrogen in biological reactions. They occur in numerous biochemical pathways, but they are of particular importance in AEROBIC RESPIRATION.

deliquescence when a substance picks up moisture from the air and may eventually liquefy.

denaturation the disruption of the weak bonds that hold a PROTEIN together, usually caused by extreme heat, or addition of a strong acid or alkali. Most denatured proteins precipitate and cannot refold into their original structure. During cooking, the proteins within an egg become denatured, causing both the yolk and the white of the egg to solidify. Extremes of temperature are fatal to the majority of all animals when the protein molecules, called EN-ZYMES, undergo irreversible denaturation and cannot perform their essential function as CATALYSTS of the biochemical processes necessary for life.

dendrite one of numerous, thin branching exten-sions attached to the main body of a NEURON cell. The dendrites are at the receiving end of the neuron and the spreading network they create has the effect of increasing the receptive area where impulses arrive from the synaptic terminals of axons of other neu-rons (*see* SYNAPSE).

dendrochronology a dating technique which uses the growth rings of trees. The concentric rings of growing trees are dependent upon climatic condi-tions and it is possible to date the rings in living trees and then to use the pattern of rings to date fossil trees or pieces of wood, providing the respec-tive lifespans overlap. One of the species of tree used as a standard is the bristlecone pine, which can live up to 5000 years and is used to date specimens older than this.

denitrification a process that forms part of the NITROGEN CYCLE and that results in the return of molecular nitrogen to the atmosphere. Aerobic bacteria in the soil use ammonia and leave nitrates that are used by plants. However, some nitrates are used by other bacteria that obtain their metabolic oxygen from this source and not directly from molecular oxygen. As a result of this denitrification, a small amount of nitrogen is released into the atmosphere. The bacteria active in this way include the genera *Pseudomonas* and *Bacillus* (*see also* NITROGEN CYCLE).

dentine a material, similar in composition to bone, that lies between the pulp cavity and enamel and forms the greater part of a tooth. Dentine differs from bone in that it contains tiny ducts for blood capillaries, nerve fibres and the extensions of ODONTOBLASTS. The ivory of elephants' tusks is also made of dentine.

dentition the complete configuration of an animal's teeth including number, type and positions. If the teeth are similar in size and structure, the term used is HOMODONT, as in most non-mammals. HETERODONT dentition indicates a mixture of tooth size and type as found in most of the mammals. The arrangement of teeth in a mammal can be expressed in a dental formula which shows the number of each type of tooth. One half of each jaw (upper and lower) is represented by numbers on either side of a horizontal line. There are four numbers depending upon the incisors, canines, premolars and molars present. Therefore a person with a full mouth of 32 teeth is represented thus:

$$\frac{2\ 1\ 2\ 3}{2\ 1\ 2\ 3}$$

where the left-hand column is the number of incisors.

deoxyribonucleic acid *see* **DNA**.

dermis a layer of the SKIN which lies between the outer and thinner EPIDERMIS and the inner SUBCUTA-NEOUS tissue. The dermis consists of connective tissue with numerous collagen fibres traversed by blood and lymph vessels, various nerves and RECEP-TORS, sweat glands, hair follicles and CHROMATO-PHORES.

desalination the production of fresh water from sea water. A number of processes may be used including evaporation, distillation, freezing, and reverse OS-MOSIS although these tend to be costly.

desert a significant BIOME which has sporadic and very low rainfall and thus low levels of plant cover. There are both hot and cold deserts, the former exhibiting summer temperatures over 35°C and the latter including the TUNDRA and areas of constant ice and snow. The driest deserts, including the world's largest, the Sahara, may have less than 20 mm of rainfall per annum and consequently the vegetation is specially adapted to the conditions. In all but the driest areas, there are shrubs and plants that can resist drought conditions, e.g. cacti, succulents. Annual plants are characterized by a rapid cycle of growth which is speedy when water is available. The seeds of such plants survive in the soil during the driest conditions. Cacti and other succulents store water and some varieties expand to accommodate extra supply when it is available. Animal life also shows numerous adaptations whether it be behavioural, e.g. nocturnal or seasonal activity, or physiological, e.g. special mechanisms to conserve water.

desiccant a substance that absorbs moisture

(hygroscopic) and is thus used as a drying agent. Examples are calcium chloride, silica gel and phosphorus pentoxide.

detergent a soluble substance that acts as a cleaning agent and is particularly effective in the removal of grease and oils (both HYDROCARBONS). The hydrophobic part of a detergent molecule will readily interact with hydrocarbons, whereas the hydrophilic part will readily interact with water molecules. When added to an immiscible mixture of water and, say, grease, the "dual reactive" nature of the detergent brings the water and grease together, thus allowing them to be rinsed away.

Devonian the fourth of six Palaeozoic periods (*see* APPENDIX 5), which lasted from 408 to 360 million years ago. It is characterized by both marine and continental deposits, the former divided into stages based upon abundant fossil remains. The Old Red Sandstone continental deposits are noted for their fossil fish and early land plants. The marine fauna includes corals, brachiopods (*see* BRACHIOPODA), crinoids and goniatites (ammonoids, a sub-class of cephalopods). The Devonian also saw the decline of the TRILOBITES and the demise of the graptolites.

dextrorotatory compound any substance that, when in crystal or solution form, has an optically active property that rotates the plane of polarized light to the right (clockwise), e.g. the sugar D-glucose (D stands for extrorotatory).

diabetes a disorder in which the body's GLUCOSE balance (blood sugar level) is upset due to a lack of INSULIN which means cells do not take up glucose from the blood. This results in a high blood sugar level, so much so that the KIDNEY excretes glucose, hence the customary urine test. Associated conditions include excessive urination and a heavy thirst. Glucose is a major source of the body's energy thus its unavailability means that fat has to be broken down and this releases acidic metabolites into the blood, potentially a life-threatening situation. There are two forms of *Diabetes mellitus* which can be treated either by insulin injections or dietary control (*see also* PANCREAS).

dialysis the employment of a semipermeable membrane to separate large and small molecules by selective diffusion. Typically large molecules are starch, protein, and small molecules, salts, glucose and amino acids. If a mixture of molecules is placed in a semipermeable container (e.g. cellophane, a sheet cellulose) within a "bath" of distilled water, the smaller molecules diffuse through the membrane into the water, which itself must be replenished. This is the principle behind kidney dialysis machines which mimic the process in natural kidneys that remove nitrogenous wastes produced during METABOLISM.

diaphragm a sheet-like membrane of muscle and tendon (covered by a SEROUS MEMBRANE) that divides the mammalian thoracic and abdominal cavities. It is an important component in breathing, being raised up in an arch (resting position) during exhalation. During inhalation, the diaphragm flattens thus reducing pressure in the thoracic cavity, helping to draw air into the lungs.

diastase(s) an alternative name for amylases, ENZYMES that break down starch. Plants contain β- and α- amylase, whereas animals contain only the latter. Specifically, diastase is the part of malt containing β- amylase, and is thus important in the brewing industry.

diastole the point in a heart beat (or heart cycle) when the heart relaxes between contractions and the ventricles fill with blood. Systole (*see* SYSTOLIC BLOOD PRESSURE) and diastole last a little under half a second each and at the end of diastole the ventricles are about three quarters full (they are filled by contraction of the atria at the beginning of systole).

diatoms *see* **Bacillariophyta**

dichotomy dividing into two equal parts, e.g. in botany, where the growing point of a plant divides to produce two further, equal, growing points which similarly divide after more growth. In astronomy, when a planet is seen half-illuminated, e.g. the Moon.

dicotyledon *see* **monocotyledon**.

diffusion the natural process by which molecules will disperse evenly throughout a particular substance. The molecules will always travel down their concentration gradient; that is, they will move from a region where they are highly concentrated to a region where they are lower in concentration within that substance. Diffusion occurs in gases, liquids and, depending on the size of the molecules, across the membranes of cells and ORGANELLEs within cells.

digestion the process of breaking down food into small organic compounds that an organism can use and readily absorb. It may occur outside the organism, as with SAPROTROPHS, or within a specialized organ internally, e.g. the gut (*see* ALIMENTARY CANAL). In some cases special cells (PHAGOCYTES) engulf food particles and break them down intracellularly (*see also* STOMACH, DUODENUM and ILEUM).

dimorphism the term applied to the existence of two clearly distinct types of individual within a species. The commonest occurrence is that of sexual dimorphism where pronounced differences occur between male and female. Secondary sex characteristics are involved, e.g. plumage in birds, antlers on deer, differences in size, and the male usually shows the greater degree of adornment. It is not exclusively a feature of animals, and some plants exhibit differences in, for example, leaf shape.

dinosaur any of a large group of terrestrial reptiles of great diversity that were dominant during the JURASSIC and CRETACEOUS periods, some 190 to 65 million years ago. The two orders were the Ornithischia (herbivorous) and the Saurischia (mainly carnivorous). In addition, many of the former were heavily armoured quadrupeds, while the latter contains the largest known carnivore *Tyrannosaurus* and the herbivore *Brontosaurus*.

The dinosaurs exhibited a tremendous variety of body shape and size, and included flying forms (*Pterosaurs*). Although the general view is that in the main these were all slow, cold-blooded creatures, there is evidence to suggest that many were fast-moving and endothermic, i.e. they kept their bodies warm by metabolic reactions.

The climate in the Cretaceous period changed considerably, and for reasons that are still not absolutely clear the vast majority of the dinosaurs disappeared in an unprecedented mass extinction occurring over five to ten million years at the end of the Cretaceous.

diploid a term used to describe a cell having two of each CHROMOSOME in its nucleus. All diploid organisms will have HOMOLOGOUS pairs of chromosomes, with each member having a similar distinctive shape. The homologous chromosomes of a pair contain GENES that code for the same products during PROTEIN synthesis, although the information could result in a different form of protein and thus a characteristic (*see* ALLELE). Humans are diploid, with each cell of the body (except GAMETES) having 46 chromosomes; that is, 22 pairs of homologous AUTOSOMES and a pair of sex chromosomes.

Diptera a large insect order with approximately 80,000 species. Although these are called the true or two-winged flies, the back pair of wings is modified to form the halteres, which are used for balance. The fly's mouthpart can pierce and suck or lap. Members of the order, dipterans, feed on sap, nectar and decaying organic matter. Certain forms, e.g. the mosquitoes, feed on blood. They have a larval stage (maggot) and metamorphose into the adult form via a pupal stage.

dissociation the process by which a compound breaks up into smaller MOLECULES, or IONS and ATOMS.

dissolution the dissolving of a substance in a liquid to form a homogeneous solution.

diurnal daily—recurring every 24 hours (*see* CIRCADIAN).

dizygote a term used to describe twins who have developed from two separate ova and therefore have different genetic characters (*see also* FRATERNAL TWINS).

DNA (*abbreviation for* **deoxyribonucleic acid**) a NUCLEIC ACID and the main constituent of chromosomes. Made up of a double helix of two long chains of linked NUCLEOTIDES, BASE PAIRED between ADENINE and THYMINE or CYTOSINE and GUANINE, it transmits genetic information in the form of GENES from parents to offspring.

DNA polymerase an enzyme that helps in the replication of new DNA strands by using the uncoiled double-helix of DNA as a template on which to add free NUCLEOTIDES. Three DNA polymerases have been identified, and all of these enzymes help catalyse the step-by-step addition of a DNA nucleotide to the end of a growing DNA chain:

DNA polymerase I—gap filling and repair enzyme.
DNA polymerase II—function unknown at present.
DNA polymerase III—responsible for the majority of DNA synthesis

dominance a genetic concept that a certain ALLELE of a gene will mask the expression of another allele known as the recessive. The PHENOTYPE of the individual is a result of the expression of the dominant allele with the recessive allele (*see* HETEROZYGOTE, HOMOZYGOTE) remaining undetected as it will have no effect on the phenotype. Dominance in animal behaviour is when one individual gains priority over others of the group because of its previous success in any contest to gain superiority. It may apply to mating, food and the less dominant members often exhibit appeasement behaviour to lessen aggression. Appeasement may consist of adopting a certain posture such that the body is in a vulnerable position. Within a group a hierarchial structure may develop, with each member occupying a certain position below some, but above other individuals in the group. Because this effect was first observed with domestic fowl, it has been called the *peck* or *pecking order*.

dorsal the surface of a plant or animal which is

uppermost, i.e furthest from the ground or support. In vertebrate animals the dorsal surface is that along which the backbone runs, and in upright animals such as humans it equates with the POSTERIOR surface.

double bond the linking of two atoms through two COVALENT BONDS in a compound, i.e. the sharing of two pairs of electrons.

Down's syndrome a syndrome created by a congenital chromosomal disorder, which manifests itself as an extra chromosome 21, making 47 in each body cell. The result is the production of characteristic facial features—a shorter, broader face with slanted eyes (similar to the Mongolian races; hence the old term of "mongolism"); shorter stature, weak muscles and the possibility of heart defects and respiratory problems. The syndrome also confers mental retardation.

Down's syndrome occurs approximately once in every 700 live births, and although life expectancy is reduced individuals can live beyond middle age. However, the development of leukaemia and Alzheimer's disease is more likely for sufferers of Down's syndrome. An additional effect is that most individuals are sexually underdeveloped.

The incidence of the syndrome increases with the age of the mother from 0.04 per cent of children for women under 30 to 3 per cent for women at 45. Above 35, therefore, it is likely that an AMNIOCENTESIS test would be undertaken.

Drosophila a fruit fly from the order Diptera which is greatly used in genetic research. Fruit flies breed quickly and easily and have just four pairs of chromosomes that can be viewed under the microscope.

It is thus used to study LINKAGE and cytogenetics (inheritance related to cell structure and function).

duodenum the first part of the vertebrate's SMALL INTESTINE where food (CHYME) from the stomach is acted upon by BILE and pancreatic enzymes. Also active are various enzymes from the duodenum itself, e.g. SECRETIN (*see also* SUCCUS ENTERICUS), that assist in the breakdown of fats, proteins and carbohydrates. The duodenum neutralizes the acidic stomach secretions producing the alkaline conditions required for the intestinal enzymes. The duodenum is about 25 cm long.

E

ear the vertebrate sense organ used for detection of sound and the preservation of balance. It comprises three parts, the outer, MIDDLE and INNER EAR, the first two acting to collect sound waves and transmit them to the inner ear, which contains the hearing organs and balance mechanism. Ear is also used to describe the external skin and cartilage of the outer ear as seen in mammals.

eardrum *see* **tympanum**

ear ossicles three bones in the MIDDLE EAR of mammals; the INCUS, MALLEUS and STAPES (or anvil, hammer and stirrup, respectively). The ossicles bridge the middle ear and carry vibrations to the fenestra ovalis (oval window) into the INNER EAR. In amphibia and some reptiles there is a single bone, the *columella*, that carries vibrations from the eardrum to the oval window.

ecdysis (*plural* **ecdyses**) the process whereby arthropods (insects and crustaceans) undergo shedding of their old and secretion of a new exoskeleton, to permit growth. The moult is activated by ecdysone, a STEROID HORMONE, which also prompts the synthesis of proteins required for these changes. Ecdysis commences with the resorption of some materials from inside the old CUTICLE (synonymous with

EXOSKELETON in arthropods) and the formation of the new skeleton, which, to begin with, is soft. The old exoskeleton splits and is shed and the animal increases its size (by taking in air or water) while the new exoskeleton hardens under the agency of CHITIN and in some cases (e.g. crabs) calcium salts.

ECG (*abbreviation for* **electrocardiograph**) equipment used to record the current and voltage associated with contractions of the heart.

Echinodermata a phylum containing marine invertebrates, which in its classes includes the starfish, brittle stars, sea urchins, sea cucumbers, crinoids or sea lilies and the recently discovered sea daisies. A number of features are common to these creatures: a pentaradial symmetry in the adult form, which commonly is shown as five arms radiating from a central disc or body; tube feet "powered" by a water VASCULAR SYSTEM; an EXOSKELETON (or *test*) of calcareous plates and in many cases bumps or spines.

The canals of the water vascular system branch into the thousands of tube feet that are used for movement and also feeding and respiration. Each tube foot is an evagination of the body wall, which acts like a suction pad and, in concert with the other feet, enables the creature to hold on to the substrate or to grasp prey. Sea urchins do not have arms but five rows of tube feet allow a slow movement. Some echinoderms can regenerate lost arms. The nervous system comprises a few nerve cords and the COELOM is well developed. This phylum has existed for a very long time, as fossil forms have been found in rocks 500 million years old.

echo location the use of high pitched pulses of sound to determine the location of objects. Nocturnal animals such as bats and aquatic creatures, e.g. dolphins, emit sound and receive the reflected signal, which enables them to determine the position and distance of an object.

ecology the study of the relationship between plants and animals and their environment. Ecology is also known as bionomics and is concerned with, for example, predator-prey relationships, population dynamics and competition between species.

ecosystem the sum total of biological life and non-biological components within an area and their interaction and inter-relation. Thus an ecosystem includes all the organisms in a community and the geology, chemistry and climate, and may be a lake, a forest, a region of tundra, etc. There are numerous cyclical processes including energy, which starts as sunlight, is converted to chemical energy by the primary producers or autotrophs (i.e. green plants, *see* AUTOTROPHISM) is consumed by heterotrophs (animals) and released as heat. Nutrients are returned to the system as wastes (*see* NITROGEN CYCLE, *also* CARBON CYCLE).

Many ecosystems are under considerable threat from man, whether due to agriculture, industry, commercial developments, warfare or accidents such as oil spills.

ecotype a population that is genetically adapted to the local conditions of its particular habitat.

ectoderm the outer layer of cells on the GASTRULA, which eventually give rise to the EPIDERMIS, and, in some phyla, to the CENTRAL NERVOUS SYSTEM (*see also* GERM LAYERS).

ectoparasite a parasite that lives on the surface of its host. In the case of man ectoparasites include

fleas, mosquitoes and lice (*see also* PARASITE and
ENDOPARASITE).

Ectoprocta *or* **Bryozoa** ("moss animals") a phylum
of mainly marine coelomates (*see* COELOM), which
includes the "sea mats" as well as moss animals.
Individual animals are tiny (approximately one
millimetre across) but they form colonies reaching
half a metre or more. Many varieties have a hard
EXOSKELETON through which the LOPHOPHORE extends.
Of the five thousand or so species, many are important as reef builders. The individuals resemble
COELENTERATE POLYPS, and they feed by microphagy.

EDTA (*abbreviation for* ethylene diamine tetra-ace-
tic acid $CH_2N(CH_2COOH)_2$) a chelating (*see* CHELA-
TION) agent often used in TITRATION of metal ions.
Because of its readiness to combine with metals, it
can be used to protect enzymes from inhibition by
metals, i.e. the prevention of normal enzyme activity, because the inhibitor binds to the active site on
the enzyme, thus blocking the substrate. EDTA can
also be used with some culture media as a slow
release agent for metal ions.

EEG *abbreviation for* ELECTROENCEPHALOGRAM.

effector when an animal receives a stimulus, whether
internal or external, the effector is the cell or organ
that produces the physiological response. In man,
muscles and glands are examples.

efferent the term applied in the main to nerves or
blood vessels that are carrying impulses or blood
respectively away from the appropriate organ, e.g.
in this case it could be from the central nervous
system and from the gills.

Elasmobranchii a sub-class of CHONDRICHTHYES

electrocardiograph *see* **ECG**.

electrode a conductor that facilitates passage of an electric current into or out of a liquid (as in ELECTROLYSIS) or a gas (gas tube) or a vacuum (valve). (*See also* CATHODE and ANODE.)

electroencephalogram (EEG) a record of the brain's electrical activity. By placing electrodes on the scalp, the electrical waves from different parts of the brain can be recorded. There are four main types of waves: alpha, beta, theta and delta. Alpha waves are produced when awake, and delta waves occur in sleeping adults. Beta waves occur when there is mental activity and children or adults under stress produce theta waves. The occurrence of delta waves in wakeful adults indicates brain damage and detecting such disorders is one use of an EEG.

electrolysis chemical decomposition achieved by passing an electric current through a substance in solution or molten form. IONS are created, which move to electrodes of opposite charge where they are liberated or undergo reaction.

electrolyte a compound that dissolves in water to produce a SOLUTION that can conduct an electrical charge.

electron an indivisible particle that is negatively charged and free to orbit the positively charged NUCLEUS of every atom. In the traditional model, electrons move around in concentric shells. However, the latest concept, based on quantum mechanics, regards the electron moving around the nucleus in clouds that can assume various shapes, such as a dumb-bell (two electrons moving) or clover leaf (four moving electrons). The shape and density of the outermost electronic shell will help determine what reactions are possible between particular atoms and molecules, e.g. whether an atom will easily gain or lose electrons to form an ION.

electronegativity a measure of the power of an atom within a MOLECULE to attract electrons. Every element in the PERIODIC TABLE is given an electronegativity rating based on an arbitrary scale in which fluorine is given the highest rating of 4, as it has the most electronegative atoms. Electronegativity differences between the different atoms within a molecule can be used to estimate the nature of the bonds formed between those atoms, i.e. whether it is a COVALENT, IONIC or POLAR COVALENT BOND.

electron microscope a microscope which uses a beam of electrons rather than a beam of light impinging upon the object. This gives much greater resolution. In the *transmission* electron microscope the electron beam passes through a very thin slice of the object and an image is created due to the scattering of the beam which is focused and enlarged dramatically onto a fluorescent screen. *Scanning* electron microscopy involves scanning the surface of a sample and the image is generated by secondary electrons. The magnification of the object is less with scanning electron microscopy but a three-dimensional image is created.

electron pair two electrons from the outer shells of two atoms which are shared by the adjacent nuclei to form a bond.

electron transport chain *or* **respiratory chain** a chain of reduction or oxidation reactions occurring in PHOTOSYNTHESIS and AEROBIC respiration, which facilitates the passage of electrons along a sequence of proteins. The stepwise nature of the process stores energy from the individual reactions in such a way that it can be used to generate ATP. The participating molecules are in the MITOCHONDRION and the whole process links with GLYCOLYSIS outside the mitochondria of cells and the KREBS CYCLE within the cell, accounting ultimately for most of the ATP that is produced by cellular respiration.

electrophoresis a method for separating the molecules within a solution using an electric field. The molecules will move at a speed determined by the ratio of their charge to their mass. All electrophoretic separations involve placing the mixture to be separated onto a porous, supporting medium, such as filter paper, or a plant-derived gel called agarose that has been soaked in a suitable BUFFER. An electric field is applied, causing the molecules, now dissolved in the conducting buffer solution, to move at a rate according to their charge-to-mass ratio. The various molecules can be identified by comparing their final position on the supporting medium with the position of known standards. Gel electrophoresis uses a synthetic polymer, such as polyacrylamide, as the supporting medium, and can be used to separate the different lengths of DNA chains found within the cellular nucleus of an individual. This makes gel electrophoresis an essential technique of "genetic fingerprinting," as the separated DNA strands will contain genes unique to a particular person, thus helping in identification.

element a pure substance that is incapable of separating into a simpler, different substance when subjected to ordinary chemical reactions. There are 105 elements known to us, but only 93 of these occur naturally; the others have been created in laboratories. The elements are classified into the PERIODIC encephalin

TABLE, according to the number of protons in their nucleus, i.e. their atomic number. Each element of the periodic table consists of atoms with a unique number of protons in their nuclei.

ELISA (*abbreviation for* enzyme linked immuno-sorbent assay) an IMMUNOASSAY technique that utilizes a colour change generated by a reaction catalysed by enzymes to measure the amount of a protein (or ANTIGEN) in a sample. A measured sample is put on a substrate that also contains an ANTIBODY specific to the protein being tested so that the two are bound together. A second antibody is added, which binds to a second site on the protein, and this is linked to an enzyme that acts to produce a colour change in a fourth component added to the sample. The resulting colour change is measured to give the concentration of the protein in the original sample.

embryo the developmental stage of animals and plants that immediately follows fertilization of the egg cell (ovum) until the young hatches or is born. In animals, the embryo either exists in an egg outside the body of the mother or, as in mammals, is fed and protected within the uterus of the mother.

enamel the outer covering of teeth above the gum. Enamel is a smooth, white, hard material comprising calcium phosphate, calcium carbonate and small amounts of organic compounds. Enamel is also found in certain scales of some fish.

encephalin a PEPTIDE, consisting of a short chain of amino acids, that acts as a NEUROTRANSMITTER. Two encephalins have been identified, both binding to opiate receptors in the brain and thus functioning as analgesics when their release acts to control pain. It would seem that the action of ENDORPHINS (natural painkillers) is copied. Encephalins have also been found in the nerve cells of the spinal cord.

endocrine system the network of glands that secrete signalling substances (HORMONES) directly into the bloodstream. This enables the secreted hormone to travel to, and thus affect, distant target cells, as opposed to just affecting cells that surround the endocrine gland. The PITUITARY GLAND (at the back of the head, base of the brain), the thyroid gland (in the neck) and the adrenal glands (above both KIDNEYS) are three of the major endocrine glands within the human body. Each secretes a different hormone that will subsequently affect different parts of the body. For example, the pituitary gland secretes growth hormone, the THYROID gland secretes thyroxine, and the adrenal glands secrete the stress hormone called GLUCOCORTICOIDS, all of which control various functions in the body.

endoderm the inner layer of cells on the GASTRULA that eventually give rise to (in mammals) the EPITHELIUM of the alimentary canal and respiratory tract, digestive glands (liver, pancreas) and other glands such as the thyroid (*see also* GERM LAYERS).

endodermis in the structure of a root, the layer (one cell thick) that encircles the STELE and lies at the inner limit of the CORTEX. The endodermis acts as a barrier permitting the passage of certain substances from the soil solution into the tissue of the stele.

endoparasite (*see also* PARASITE) a parasite that lives in the body of the "host." Endoparasites tend to be highly specialized, often with complex life cycles involving more than one host. The specialization may take the form of a resistant outer CUTICLE to permit existence in the acids of an animal's gut, or hooks to facilitate holding onto tissues. Many parasites create illness in the host; for example, tapeworms may cause intestinal blockage and nutritional deficiencies, and flukes (*see* TREMATODA) can cause anaemia and dysentery.

The success of such parasites lies not only in their specialization, but also in the production of vast numbers of eggs, which secures the survival of a proportion of individuals.

endopeptidase an enzyme such as pepsin which acts in PROTEOLYSIS by hydrolysing (*see* HYDROLYSIS) certain PEPTIDE BONDS in a PROTEIN.

endoplasmic reticulum a network of membrane-bound tubules and flattened sacs connected to the membrane of the nucleus found within all EUCARYOTIC cells. If the endoplasmic reticulum (ER) has RIBOSOMES attached to the outside membrane, then it is called "rough ER," but in the absence of ribosomes it is called "smooth ER," or GOLGI APPARATUS. Both the rough ER and Golgi apparatus have functions important in the biosynthesis of other organelles and in the synthesis, modification and sorting of proteins.

endopterygote a term describing insects that undergo complete metamorphosis, with a pupal stage, in their life cycle. This covers numerous orders including beetles, the true flies (DIPTERA), butterflies and moths (LEPIDOPTERA) and the Hymenoptera (bees, ants and wasps).

endorphin (*see also* ENCEPHALIN) a peptide in the PITUITARY GLAND which acts as a NEUROTRANSMITTER. Endorphins have pain relieving effects but in addition are involved in urine output, control (depression) of respiration, learning and sexual activity.

endoskeleton a framework of hard bones, spicules, plates, etc, that lie within the body of an animal. Lower animals, such as sponges and echinoderms have spicules and plates respectively but vertebrates have a skeleton of numerous bones. In addition to providing support, the endoskeleton gives protection for certain organs and points of attachment for muscles and also acts as a lever when muscles contract. The vertebrate skeleton is divided into an *axial* part—the skull, backbone and rib cage; and an *appendicular* part comprising the limbs, pectoral and pelvic girdles. The skeleton in mammals is made up of over 200 bones, some of which are joined by ligaments while others are fused together.

endosperm a tissue, formed particularly in flowering plants, that is rich in starch and other food reserves and surrounds the developing embryo in a seed. The seeds of certain plants contain a large endosperm, e.g. corn, and this and other cereal or oil crops are cultivated for this reason.

endospore a resistant cell formed by bacteria, perhaps due to unfavourable environmental conditions. A spore with a tough coat forms within the parent cell, and the latter then disintegrates, leaving the endospore, which can tolerate all sorts of extremes—heat, cold, or lack of water and food. Even boiling water may not kill the endospores of certain bacteria, and autoclaving (*see* AUTOCLAVE) is necessary (also to sterilize equipment and utensils in the laboratory). When conditions are more favourable, the endospore revives and begins growing.

endothelium a layer of cells, just one cell thick, that lines the inner surface of the heart, and the blood and lymph vessels. It is mesodermal (*see* MESODERM) in origin.

energy the capacity to do work. There are various forms of energy, including light, heat, sound, mechanical, electrical, kinetic and potential, but all are expressed in the same unit of measurement, called the JOULE (J). Energy has the capacity to change from one form to another (**energy transfer**), but the original input of energy tends to be greater than the final output during energy transfers. As the law of conservation of energy states that it is impossible to make or destroy energy, the difference in the input/output energy levels is a result of the conversion of some of the input energy into an unwanted form, e.g. heat instead of mechanical energy. The energy content of a system or object can be regarded as the "work done" by it and can be calculated using the following equation:

Work Done (w) = Force (F) x Distance Moved (S)

enthalpy the quantity of heat energy (THERMODYNAMICS) possessed by a substance. Enthalpy (H) has units of joules per mole ($Jmol^{-1}$) and is defined by the equation:

$H = E + PV$ where E = internal energy of a system

P = pressure

V = volume

The enthalpy change (δH) during a reaction is referred to as δH negative when the reaction is exothermic (heat evolved) and δH positive when the reaction is endothermic (heat absorbed).

entomology the study of insects.

entropy a measure of the randomness (disorder) of a system. It is a natural tendency of the whole universe that allows all energy to be distributed.

Entropy (S) has units of joules per kelvin per mole (JK^{-1} mol^{-1}) and can be related to ENTHALPY (H) using Gibb's law ($G = H - TS$). The greater the disorder, the higher the value for entropy, but at absolute zero entropy is also zero.

enzyme any PROTEIN molecule that acts as a natural CATALYST and is found in the bodies of all bacteria, plants and animals. Enzymes are essential for life as they allow the complex chemical reactions of biochemical processes to occur at the relatively low temperature of the body. Enzymes are highly specific in that they will only act on certain SUBSTRATES at a specific pH and temperature. For example, the digestive enzymes called amylase, LIPASE and trypsin will only work in alkaline conditions (pH > 7), whereas the digestive enzyme, pepsin will only work in acidic conditions (pH < 7). Enzymes operate through the provision of ACTIVE SITES (there may be one or more per enzyme) which substrate molecules can bind to, thus forming an intermediate (or enzyme-substrate complex) that lasts for a very short time. However, this increases the reaction rate dramatically and product is formed whereupon the active site is freed. Some enzymes bind only one substrate molecule, others bind a variety. The names of most enzymes end in *-ase*, and this is added onto the name of the substrate, thus a peptidase breaks down peptides.

Eocene a Tertiary epoch that began at the end of the Palaeocene and ended at the Oligocene (from 54 to 38 million years ago). It is particularly noted for the mammalian expansion, notably the horses, whales and bats, which all appeared, and the presence of rodents.

eon the largest geological unit of time (*see* APPENDIX 5), e.g. Phanerozoic, which includes the Palaeozoic, Mesozoic and Cenozoic eras.

ephemeral an animal that is short-lived (e.g. the mayflies, the order *ephemeroptera*) or a plant that has a short life cycle and thus more than one generation in one year, e.g. willowherb, groundsel.

epidermis the outer layer of cells that cover a plant, providing protection and reducing water loss. Depending upon the specific part of the plant, the epidermis may have other functions, such as the secretion of a wax-like coating—the cuticle—over leaves and stems.

In animals, the epidermis performs a similar function. Invertebrates have an epidermis beneath an impermeable CUTICLE. Vertebrate skin is made up of the epidermis over the thicker DERMIS, and the former is composed of several layers, the outer one (stratum corneum) providing a water-resistant covering.

epididymis a coiled tube in which sperm from the testes are stored and where they mature prior to ejaculation via the VAS DEFERENS.

epiglottis a flap made of cartilage attached to the pharynx wall, which covers the windpipe when food is swallowed.

epilepsy a seizure disorder caused by lesions in the brain. The symptoms are in the form of attacks, known as fits, that can include a feeling of numbness, muscular convulsions, inability to speak, etc. Epilepsy can be controlled by certain drugs, but in bygone days, surgery was performed on patients who suffered frequent and extreme attacks in an attempt to control these. The operation disconnected the left and right hemispheres of the brain by cutting the communication system between them, a fibrous network called the *corpus callosum*.

epiphyte a plant that grows upon another plant purely for support although it, the former, is not rooted in the ground. The commonest examples are mosses, lichens and some tropical orchids.

epithelium (*plural* **epithelia**) a tissue that is composed of tightly packed cells and covers the outer surface of the body and lines the wall of cavities in the body. The epithelium provides a barrier against injury, micro-organisms and, to varying extents, fluid loss. One surface of the epithelium is thus free and the other fixed to a basement membrane. Epithelia, which may be of endodermal or ectodermal origin (*see* ENDODERM and ECTODERM respectively), are distinguished by cell shape and layers. Simple and stratified epithelium refers to single and multiple cell layers; cuboidal, columnar and squamous (flat pads) refer to shape. The shape fits the function, for example, simple squamous epithelia permit diffusion and line the air sacs of the lung, while columnar epithelia, which can secrete solutions and absorb nutrients, line the stomach and intestines.

ER *see* **endoplasmic reticulum**.

erythroblast cells in the red bone marrow that develop into red blood cells. They begin colourless and nucleated but fill with haemoglobin and in mammals they lose the nucleus.

erythrocyte the red blood cell of vertebrates that is made in the bone marrow. It differs from other cells in the human body in that just before it is released into the bloodstream it sheds its nucleus. Erythrocytes contain HAEMOGLOBIN, the protein molecule essential for transportation of oxygen from the lungs to all tissues in the body. Within the erythrocyte membrane there is a complex molecule called the Band 3 protein, which is essential for the transport of carbon dioxide (CO_2) from all body tissue to the lungs as it allows CO_2, in the form of the bicarbonate ion (HCO_3^-), to leave the red blood cell in exchange for a chloride ion (Cl^-). A person deficient in the number of erythrocytes circulating in their blood is said to be anaemic. Anaemia can be a result of a defect in the structure of the erythrocyte membrane or lack of iron to form the haem group of the haemoglobin molecule.

essential amino acid of the twenty amino acids required, eight or nine are "essential," that is, must be in the diet because they cannot be synthesized. A lack will lead to protein deficiency. The essential amino acids are available in meat, cheese and eggs, but if culture, environment or any other factors limit such supply, a balanced vegetarian diet will supply the correct compounds, for example corn and beans together provide all eight amino acids. Since amino acids cannot be stored in the body, the supply must be regular otherwise synthesis of proteins will be affected.

essential element certain elements are necessary for organisms, albeit in small amounts, to ensure good growth and development. In addition to those found in organic compounds, major elements required include calcium, phosphorus, sodium, potassium, chlorine, magnesium and sulphur. Trace elements, which are required in minute quantities, include manganese, iron, cobalt, copper, and zinc. Many of the trace elements are components of, or are associated with, enzymes or complex molecules, while the major elements may be important in building the body, such as magnesium, calcium and phosphorus in bone, or in forming body fluids, as with sodium and potassium.

essential fatty acid most of the fatty acids required can be produced in the body but certain UNSATURATED fatty acids must form part of the diet, e.g. linoleic acid, which is necessary for the production of some membrane PHOSPHOLIPIDS.

essential oil scented oils secreted by some aromatic plants. The oils, which are composed mainly of TERPENES, are extracted for use in flavourings, perfumes and medicines, e.g. flower oils, oil of cloves.

ester an organic compound formed from acids by replacing the hydrogen with an alkyl radical, e.g. CH_3COOH (ethanoic acid) to $CH_3COOC_2H_5$. Many esters have a fruity smell and are used for flavourings. Esters are common in nature, as animal fats and vegetable oils are formed from mixtures of esters.

estuary a partially enclosed stretch of water that is subjected to marine tides and fresh water draining from the land. An estuary is usually created as a drowned valley due to a post-glacial rise in sea level. A large amount of sediment is deposited in estuaries and the tidal currents may produce channels, sandbanks and sand waves.

ethaneidoic acid *see* **oxalic acid**.

ethanoic acid *see* **acetic acid**.

ethanol a derivative of ethane, with a functional hydroxyl group (OH) in place of one hydrogen atom, i.e. CH_3CH_2OH. Ethanol is obtained either by fermentation of carbohydrates to form alcoholic beverages, or commercially prepared from ETHENE by adding water and sulphuric acid.

ethene the first member of the ALKENE family, it is an insoluble gas at room temperature. Many plants produce ethene as a hormone or GROWTH SUBSTANCE. The gas promotes ripening of fruit and also the thickening of tissues at the expense of their elongation. It is now used to ripen fruit artificially.

ethology the precursor to behavioural ecology, which dealt with the study of animal behaviour in the context of the animal's natural environment and how this was linked to its ongoing survival.

eucaryote any member of a class of living organisms (except viruses) that has a membrane-bound nucleus within its cells. All eucaryotic cells contain ORGANELLES, which are also bound by closed, phospholipid membranes, e.g. chloroplast, endoplasmic reticulum, mitochondrion, etc. All plants and animals are eucaryotes, but bacteria and cyanobacteria are PROCARYOTES.

eugenics the study of how the inherited characteristics of a human population can be improved by genetics, i.e. controlled breeding.

Eustachian tube in vertebrates, the tube connecting the MIDDLE EAR to the pharynx. Although normally closed, it can open to permit air pressure on either side of the eardrum to be equalized. It was named after the 16th century anatomist Eustachio.

Eutheria a subclass of mammals. These are the placental mammals, where an embryo is retained in the mother's uterus and receives nutrients and excretes waste via the placenta. The group includes the following orders: Artiodactyla (deer and pigs), CARNIVORA (cats and dogs), INSECTIVORA (hedgehogs),

CETACEA (whales), RODENTIA (rats, squirrels, beavers), CHIROPTERA (bats) and PRIMATES (monkeys, apes, humans). It is thought placental and marsupial mammals sprang from a common ancestor around 100 million years ago, and then, during the late Cretaceous and early Tertiary, the group developed through ADAPTIVE RADIATION.

eutrophic a term applied to a body of water, such as a lake, that has an abundant supply of nutrients and consequently high photosynthetic production of organic matter. Eutrophication is the process of pollution of a lake due to excessive quantities of sewage or fertilizers. The excess nitrogen and phosphorus makes for rapid growth of algae, which ultimately depletes the oxygen in the lake when bacteria decompose dead algae. The final outcome is the death of fish and other animals in the lake.

evergreen a plant that retains its leaves all year, in contrast with DECIDUOUS plants.

evolution the process by which an organism changes and thus attains characteristics distinct from existent relatives. Any species of organism will only evolve if:

(a) There has been genetic mutation allowing variation in the genetic information the parent passes on to its descendants.

(b) An individual proves to be more suitable to a particular environment than its relatives, allowing it to survive and propagate whereas its relatives will become extinct, i.e. NATURAL SELECTION (*see also* DARWINISM).

excretion the process whereby an organism removes metabolic waste products from its body. The simplest method involves diffusion through the body or via specially enlarged surfaces such as GILLS. In higher animals, special systems and organs are used, e.g. the KIDNEY in vertebrates for excretion of nitrogenous wastes and the LUNGS for carbon dioxide and water.

exocrine gland a gland that is derived from epithelial cells in the embryo and secretes, usually through a duct, into a body cavity or onto the body surface.

exopeptidase an ENZYME involved in PROTEOLYSIS through the removal of single amino acids from the end of a protein molecule.

exopterygote a term describing insects that do not undergo complete metamorphosis. In their development, a series of larvae (nymphs) become more like the adult, but there is no pupal stage. Included are the ODONATA (dragonflies and damselflies), Phasmida (stick insects), Isoptera (termites) and the Hemiptera (aphids, cicadas, leaf hoppers).

exoskeleton a hard protective layer, on the surface of an animal, that also provides support and attachments for muscles. Some animals add to an outer edge to accommodate growth (e.g. molluscs and clams) while others shed their exoskeleton and replace it with a new, larger skeleton, to permit growth in spurts (*see* ECDYSIS). Exoskeletons may be calcareous (e.g. molluscs), CHITINOUS (e.g. ARTHROPODS) or in the case of crabs and lobsters the chitin is further hardened by calcium salts.

extinction the end of a particular species or lineage. Ordinarily, this could occur through "natural" agencies such as predation, disease, environmental changes, but there have been periods of mass extinction in Earth history, notably around 65 million years ago in the early TERTIARY. There is evidence to suggest that there was atmosphere cooling caused by volcanic activity and/or asteroid impact, which probably led to acid rain and temperature reduction. Vast numbers of species disappeared, including the DINOSAURS and about three quarters of marine species plus some terrestrial plants.

eye the organ responsible for sight, which varies from the complex vertebrate eye to the primitive eyespots of unicellular organisms. Eyespots are areas of light-sensitive pigments (CAROTENOIDS) that prompt the cell to move in relation to the light. ARTHROPODS have COMPOUND EYES and less complex OCELLI. The eyes of some cephalopods (squid, octopus) are similar to the vertebrate eye, which is near spherical and fluid filled. Light enters through the CORNEA and a small space filled with AQUEOUS HUMOUR before passing through the lens and the pupil in the IRIS into a larger area of VITREOUS HUMOUR. The lens focuses the image on the RETINA, which contains special cells (*see* CONES and ROD CELLS), and the impulses thus produced are transmitted to the brain along the optic nerve.

F

factor VIII (antihaemophilic factor) a soluble PRO-TEIN and CLOTTING FACTOR vital in the process of BLOOD CLOTTING. Factor VIII stimulates Factor IXa to activate Factor X, which converts prothrombin to thrombin, the ENZYME that causes FIBRINOGEN to clot the blood. Haemophiliacs lack this factor, but it can be administered and it is now produced by genetic engineering, thus eliminating the risk of contamination with viruses.

FAD (*abbreviation for* flavin adenine dinucleotide) a compound that is important in biochemical reactions. FAD is a coenzyme, i.e. a carrier molecule that functions with a certain enzyme. It comprises riboflavin (vitamin B_2), which is made up of ribose and flavin (a triple ring structure), with a phosphate group which altogether joins with AMP. FAD acts as a hydrogen carrier, being reduced to $FADH_2$, which is subsequently oxidized back to FAD in the ELECTRON TRANSPORT CHAIN.

Fahrenheit (F) a temperature scale, devised by the German physicist Gabriel Fahrenheit (1686–1736), that set the freezing point of water at 32° and the boiling point at 212°. Fahrenheit temperatures can be converted to CELSIUS by the equation $F = 1.8C + 32$.

fallopian tube *or* **oviduct** a tube with a funnel-shaped opening that conveys egg cells to the womb from the OVARY. Muscular and ciliary action move the eggs.

fall-out radioactive material deposited on the ground from the atmosphere. The source of the radioactive substances is from nuclear explosions or their escape from nuclear reactors whether through failure to adequately filter coolants or an accident.

family a taxonomic unit between ORDER and GENUS where a number of similar families are grouped into an order. The names of families usually end in *-idae* in zoology and *-aceae* in botany.

FAP *abbreviation for* fixed action pattern (*see* INSTINCT).

fats a group of naturally existing lipids that occur widely in plants and animals and serve as long-term energy stores. A fat consists of a GLYCEROL molecule and three FATTY ACID molecules, collectively known as a triglyceride, which is formed during a condensation reaction (water is released). Fats are important as energy-storing molecules since they have twice the calorific value of carbohydrates. In addition, they insulate the body against heat loss and provide it with cushioning, which helps protect against damage. In mammals, a layer of fat is deposited beneath the skin (subcutaneous fat) and deep within the tissues (adipose tissue) and is solid at body temperature due to the high degree of saturation. In plants and fish, the fatty acids are generally less saturated and as such tend to have a liquid-like consistency, i.e. oils, at room temperature.

fatty acids a class of organic compounds, containing

136

a long hydrophobic (water insoluble) hydrocarbon chain and a terminal carboxylic acid group (COOH), that is extremely hydrophilic (water soluble). The chain length ranges from one carbon atom (HCOOH; methanoic acid) to nearly thirty carbon atoms, and the chains may be SATURATED or UNSATURATED. As chain length increases, melting points are raised and water solubility decreases. However, both unsaturation and chain branching tend to lower melting points. Fatty acids have three major physiological roles:

(1) They are building blocks of phospholipids (lipids containing phosphate) and glycolipids (lipids containing carbohydrate). These molecules are important components of biological membranes, creating a lipid bilayer which is the structural basis of all cell membranes.

(2) Fatty acid derivatives serve as hormones and intracellular messengers.

(3) Fatty acids serve as fuel molecules. They are stored in the CYTOPLASM of many cells in the form of triglycerides (three fatty acid molecules joined to a glycerol molecule) and are degraded, as required, in various energy-yielding reactions.

feather (*see also* AVES) an epidermal outgrowth of KERATIN, which forms the body covering of birds and provides insulation, shape and aerodynamic properties. Each feather comprises a central shaft (rachis), which is hollow. Vanes radiate from the shaft and consist of BARBS and smaller barbules. Contour feathers provide the overall shape of the bird and barbules on these feathers also have hooks to interlock with other feathers. However, down feathers do not have these hooks, and their more random arrangement produces the typical appearance of down with its insulating qualities.

feedback mechanism a control mechanism that uses the products of a process to regulate that process by activating or repressing it. Almost all homeostatic mechanisms (*see* HOMEOSTASIS) in animals operate by negative feedback, whereby a variation from the normal triggers a response that tends to oppose it. For example, it operates during hormonal release to maintain steady blood sugar levels. Positive feedback is found less often as a biological control mechanism. Here, a variation from the normal causes that variation to be amplified, and this is usually a sign that the normal control mechanisms have broken down.

Fehling's test a test for detecting and estimating reducing SUGARS and ALDEHYDES in solution. Equal amounts of two solutions, copper sulphate, and sodium potassium tartrate with sodium hydroxide, are added to the sample, which is then boiled. A red precipitate of copper oxide indicates a positive result.

femur the thigh bone of humans and other terrestrial vertebrates. At the upper end it articulates with the hip joint and at the lower end with the TIBIA by means of two CONDYLES (*see also* JOINT).

fenestra two membranes, the *fenestra ovalis* and *fenestra rotunda* (oval and round windows) between the middle ear and inner ear. Vibrations are transmitted from the TYMPANUM to the COCHLEA via the ossicles (*see* EAR OSSICLES) and the *fenestra ovalis*. The round window reacts to changes in the perilymph pressure (*see* COCHLEA).

fermentation a form of ANAEROBIC RESPIRATION, which

converts organic substances into simpler molecules, generating energy in the process. Fermentation, carried out by certain organisms such as bacteria and YEASTS, is the conversion of sugars to alcohol in the process known as alcoholic fermentation. Lactic acid fermentation occurs in the muscles of higher animals when the oxygen requirement exceeds the supply and sugar is converted into lactic acid. In industry, fermentation is important in baking and in beer and wine production, and these use large quantities of yeast.

fertilization the fusion of male and female GAMETES to produce a single cell, which sets in motion a chain of events that gives rise to a new individual. In animals, where the gametes unite outside the parents' bodies, it is termed external fertilization (as in most fish). Where the male gametes are deposited within the body of the female by the male, it is termed internal fertilization, as is the case with mammals. In flowering plants, after pollen has been transferred from the male to the female part of the flower, a pollen tube develops, which transfers two male nuclei to the ovule of the female. Double fertilization occurs, producing a DIPLOID ZYGOTE and a triploid endosperm that act as a food supply for the developing embryo.

fertilizer a substance added to soil to replace nutrients removed by plants, thus contributing towards their health and vitality. Fertilizers may be natural (e.g. manure) or synthetic, the latter containing nitrogen, phosphorus and potassium as the main constituents.

Fibonacci series *see* **phyllotaxy**.

fibre plant fibres are elongate cells with thick walls

of LIGNIN which give support in the XYLEM. Some fibres are commercially significant, e.g. hemp. *Dietary* fibre is the part of some foods that cannot be digested, e.g. wholemeal cereals and flour, nuts and fruit have a high fibre content. Foods that are highly refined contain no fibre. There are four types of dietary fibre—cellulose, hemicellulose, lignins and PECTINS—and many consider their consumption helpful in preventing certain diseases, e.g. obesity, diverticulosis and DIABETES.

fibrin an insoluble protein formed from FIBRINOGEN, which is the framework of a blood clot.

fibrinogen a blood PROTEIN, which causes BLOOD CLOTTING due to action by the ENZYME thrombin. The end product is fibrin.

fibrinolysis the breakdown of FIBRIN by the enzyme plasmin which thus breaks down blood clots and removes the fibrin from the system.

fibula the smaller of two bones between knee and ankle in man and other terrestrial vertebrates.

filament a term commonly used to describe any long thread-like structure whether the filamentous form of ACTIN or the barbs in a FEATHER. Also, in flowering plants it refers to the stalk bearing the ANTHER which together form the STAMEN.

filter feeding a method of feeding used by animals large and small. It is very common among invertebrates where very small particles of food are taken in from the surrounding water, often through some straining device. It is also used by some large fish and mammals, e.g. certain sharks, and the baleen whale.

filter paper a pure cellulose paper used in the laboratory for the separation of solids from liquids by FILTRATION.

filtrate the liquid remaining after FILTRATION, having been separated from a solid or liquid mixture.

filtration the separation of a solid from a liquid by passing the mixture through a suitable separation medium, e.g. FILTER PAPER, which holds back the solid and permits the liquid to pass through.

fins refer primarily to the organs used by fish for locomotion and stabilization. Fins are strengthened by fin rays, which are flexible supports of cartilage or bone. Fish usually possess one or more dorsal and ventral fins for balance, a tail fin (CAUDAL) for propulsion, paired pectoral and pelvic fins for steering and stopping. These paired fins correspond to the limbs of tetrapods.

fission a form of asexual reproduction where one cell splits to form two cells as in binary fission or several cells (multiple fission). Fission is common in unicellular organisms, e.g. diatoms, protozoa (such as *Amoeba*) and bacteria. In PROCARYOTES the chromosome becomes attached to the cell membrane, duplicates, the cell grows, and the chromosomes move apart, and then a furrow divides the cell into two. Eucaryotes undergo MITOSIS followed by cytoplasmic division (cytokinesis), as with *Amoeba*.

fixation to enable study of a tissue, organism etc. in MICROSCOPY the specimen has to be killed but maintained in a state as near to natural as possible. The specimen is immersed in a chemical fixative which denatures the protein, thus preventing distortion of cells. Solutions used include formaldehyde, ethanol and osmium tetroxide and the choice depends to some extent on whether light or electron microcopy is the chosen technique.

fixed action pattern (FAP) *see* **instinct**.

flagellum (*plural* **flagella**) a long thread-like structure (up to 150µm) extending from the cell membrane of certain cells. Flagella beat in a wave-like motion, which in many instances is used for locomotion, e.g. in certain protozoans, bacteria, or in reproductive cells (e.g. sperm). In an organism that is fixed to the substrate, beating of the flagella moves the surrounding fluid.

flavonoid naturally occurring phenolic compounds that include many plant pigments. One group, the arthocyanins, form the red, blue and purple colours in flowers. Flavones are another group, conferring yellow pigmentation.

Fleming, Sir Alexander (1881–1955) a Scottish bacteriologist who discovered the antibiotic penicillin in 1928, for which he was awarded the Nobel Prize in 1945.

flower structure the flower is the reproductive part of angiosperms (i.e. flowering plants), and it varies enormously in form from the tiny flower of wind-pollinated grasses to the large bright flower that is pollinated by insects. Before the coloured PETALS open, they are enclosed by SEPALS and both play no *direct* role in reproduction. The reproductive organs are within the petals and comprise the STAMEN and CARPEL. All these parts are supported at the base by the receptacle. Flowers often form INFLORESCENCES, which may be sufficiently dense as to resemble a single flower.

fluid any substance that flows easily and alters its shape in response to outside forces. All gases and liquids are fluids. In liquids, the particles move freely but are restricted to the one mass, which occupies almost the same volume. In gases, however, the particles tend to expand to the limits of their containing space and thus do not keep the same volume.

fluid-mosaic the name given to the model that describes the structure of the cell membrane in organisms, proposed by Singer and Nicholson in the 1970s. Using electron microscopy, they confirmed that the lipid component is organized in a regular bimolecular structure with protein molecules arranged irregularly along the lipid layers. Both lipid components can move laterally in the membrane.

fluorocarbons (*see also* CFC) a group of compounds where fluorine replaces some or all of the hydrogen atoms in a hydrocarbon. The resulting compounds are similar to the original hydrocarbon in some ways, but are unreactive and thermally stable. Uses include oils and greases, fluorine-containing polymers, SURFACTANTS and also refrigerants and aerosol propellants, although these latter uses are being discontinued due to the effect of the CFCs on the OZONE LAYER.

foetus the developing young in the mammalian uterus, from the post-embryonic period until birth. In humans, this period is from about 7–8 weeks after FERTILIZATION.

folic acid a compound that forms part of the VITAMIN B complex. It is involved in the BIOSYNTHESIS of some AMINO ACIDS and is used in the treatment of anaemia.

follicle a dry fruit formed from a CARPEL, which splits when dry to release its seed(s). In anatomy, a group of cells that surround a structure or cell to provide protection, e.g. hair follicle, GRAAFIAN FOLLICLE.

food chain in its simplest terms, the route by which energy is transferred through a number of organisms by one eating another from a lower TROPHIC LEVEL. At the base of the chain are the primary producers—green plants, which are eaten by herbivores, which in turn are eaten by carnivores, and so on. In reality, the situation is much more complex, and many animals feed at more than one trophic level. The complete system of chains in a community is a food web.

foramen a natural opening in an animal's organ, etc, applied especially to bone or cartilage, which permits the passage of nerves and blood vessels.

Foraminifera a phylum of marine PROTOZOA with hard shells of calcium carbonate or silica, which form important INDEX FOSSILs and sedimentary rock sequences (chalk). Modern forms, e.g. *Globigerina*, are a component in deep sea oozes. PSEUDOPODIA protrude through the shell and function in swimming and feeding.

forebrain *or* **prosencephalon** the front section of the brain of a vertebrate embryo, which develops to form the HYPOTHALAMUS, THALAMUS and CEREBRUM in the adult.

fossil the remains of once-living plants and animals, or evidence of their existence, preserved in the strata of the earth's crust. Palaeontology, the name given to the study of fossils, has proved useful in the study of evolutionary relationships between organisms, and in the dating of geological strata.

fossil fuels NATURAL GAS, PETROLEUM (oil) and coal, today's major fuel sources. They are formed from the bodies of aquatic organisms that were buried and compressed on the bottoms of seas and swamps millions of years ago. Over time, bacterial decay and pressure converted this organic matter into fuel.

Hard coal, which is estimated to contain over 80 per cent carbon, is the oldest variety and was laid down up to 250 million years ago. Another, younger variety (bituminous coal) is estimated to contain between 45 per cent and 65 per cent carbon. The fuel values of coal are rated according to the energy liberated on combustion. Coal deposits occur in all the world's major continents, and some of the leading producer countries are the United States, China, Russia, Poland and the United Kingdom.

Natural gas consists of a mixture of HYDROCARBONS, including METHANE (85 per cent), ETHANE (about 10 per cent) and PROPANE (about 3 per cent). However, other compounds and elements may also be present, such as carbon dioxide, hydrogen sulphide, nitrogen and oxygen. Very often, natural gas is found in association with petroleum deposits. Natural gas occurs on every continent, the major reserves being found in Russia, the United States, Algeria, Canada and in counties of the Middle East.

Petroleum is an oil consisting of a mixture of HYDROCARBONS and some other elements (e.g. sulphur and nitrogen). It is called crude oil before it is refined. This is done by a process called fractional distillation, which produces four major fractions:
(1) Refinery gas, which is used both as a fuel and for making other chemicals.
(2) Gasoline, which is used for motor fuels and for making chemicals.
(3) Kerosine (paraffin oil), which is used for jet aircraft, for domestic heating and which can be further refined to produce motor fuels.
(4) Diesel oil (gas oil), which is used to fuel diesel engines.

The known residues of petroleum of commercial importance are found in Saudi Arabia, Russia, China, Kuwait, Iran, Iraq, Mexico, the United States, and a few other countries.

Together, the fossil fuels account for nearly 90 per cent of the energy consumed in the United States. As coal supplies are present in abundance compared with natural gas or petroleum, much research has gone into developing commercial methods for the production of liquid and gaseous fuels from coal.

fossilization the formation of a FOSSIL. Organisms tend to undergo some changes after death and are not usually preserved whole. In particular, the soft parts will decay and skeletal parts are often changed. Sediment may flatten an organism, and porous structures may be replaced by minerals. Recrystallization commonly occurs, replacing the fine structure of shells, and the mineral aragonite (a form of calcium carbonate) often changes to the more stable calcite. Skeletal parts may leave impressions or moulds in sediments, and these may be internal or external. In addition, burrows, trails and similar evidence of organisms can be preserved as TRACE FOSSILS.

fragmentation a type of ASEXUAL REPRODUCTION in some invertebrates, where parts of the organism break off, or the organism breaks up. The parts then develop into new individuals. It is seen particularly in the annelids (*see* ANNELIDA) and coelenterates (*see* CNIDARIA).

fraternal twins unidentical twins (dizygotic twins) that develop when two ova are fertilized simultaneously. This occurs when two ova have matured and have been shed simultaneously, and the resultant twins resemble each other only to the same extent as brothers and sisters born at different times.

freeze-drying a process used when dehydrating heat-sensitive substances (such as food and blood plasma) so that they may be preserved without being damaged during the process. The material to be preserved is frozen and placed in a vacuum. This causes a reduction in pressure, which in turn causes the ice trapped in the material to vaporize, and the water vapour can be removed, producing a dry product. For most solids, the pressure required for vaporization is quite low. However, ice has an appreciable vapour pressure, which is why snow will disappear in winter even though the temperature is too low for it to melt.

fructose a simple SUGAR ($C_6H_{12}O_6$) and STEREOISOMER of glucose, otherwise known as fruit sugar. It occurs in fruits, honey and green plants and is the sweetest of the sugars. Fructose derivatives are important in metabolic processes and fructans (polySACCHARIDE derivatives) form energy stores in some plants.

fruit a mature ovary which protects developing seeds. It comprises a fruit wall (pericarp) surrounding the seeds and while the other parts of the flower usually wither away they do develop, in some cases forming part of the fruit structure to form a *false* fruit (e.g. the apple). There are several types of fruit classified by origin, thus a *simple* fruit, e.g. the cherry, develops from a single ovary. A strawberry is an *aggregate* fruit formed from one flower with several CARPELS, and a *multiple* fruit from a tight cluster of flowers, e.g. the pineapple. A fundamental division of fruit is into dry and succulent; the former being when the ovary wall stays dry and the latter when it becomes fleshy. Usually succulents are dispersed by animals and dry fruits by wind or water.

FSH (*abbreviation for* follicle-stimulating hormone) an anterior PITUITARY GLAND secretion that in female mammals controls growth of the GRAAFIAN FOLLICLE, which produces ova. In males it stimulates formation of sperm in the TESTIS. FSH is an important component of fertility drugs.

fungus (*plural* **fungi**) simple unicellular or filamentous plants with no chlorophyll. Fungi cause decay in fabrics, timber and food, and diseases in some plants and animals. Particular fungi are used in brewing, baking, and also in the production of ANTIBIOTICS.

G

galactose a simple SUGAR ($C_6H_{12}O_6$) and STEREOISOMER of glucose. It is a constituent of LACTOSE and occurs in plant polysaccharides, e.g. gums and PECTINS.

gall bladder *see* **bile**.

gametangium a plant cell or organ that produces gametes. It is used mainly for algae and fungi.

gamete the reproductive cell of an organism. Gametes can be either male or female, and these specialized cells are HAPLOID in number but unite during fertilization, producing a DIPLOID ZYGOTE that later develops into a new organism. In higher animals, the male and female cells are called sperm and ova respectively, whereas in higher plants they are known as pollen grains and egg cells respectively. In some organisms there is essentially one type of gamete that is capable of developing into a new individual without fertilization. These gametes are usually diploid, as in the case of certain lower plant groups, e.g. many forms of algae.

gametophyte in plants which exhibit alternation of generations, it is the generation that bears the gamete-producing organs. The gametophyte is HAPLOID and in the flowering plants the male form is the pollen grain and the embryo sac is the female gametophyte (*see also* ASEXUAL REPRODUCTION).

gamma ray a type of electromagnetic radiation released during the radioactive decay of certain nuclei. The rays released are the most penetrative of all radiations, requiring about twenty millimetres of lead to stop them. The gamma rays are useful for sterilizing substances and in the treatment of cancer. They have the shortest wavelength of any wave in the electromagnetic spectrum, i.e. 10^{-10} to 10^{-12} metres.

ganglion (*plural* **ganglia**) a mass of nervous tissue that contains cell bodies (the part of the nerve cell with the nucleus) and SYNAPSES. Ganglia comprise part of the CENTRAL NERVOUS SYSTEM (CNS) in invertebrates, occurring along the nerve cords but in vertebrates they occur outside the CNS in the main.

gas the fluid state of matter capable of indefinite expansion in every direction, because of the relatively few bonds between the atoms or molecules present in the gas. If heat ENERGY is supplied to a gas, it expands to the limits of its containing vessel, exerting a pressure on this vessel that in turn exerts a force back onto the gas.

gas analysis the analysis of a mixture of gases by various means. By putting a measured quantity of gas in contact with various reagents, certain components can be measured after each phase of absorption. Carbon dioxide is absorbed by potassium hydroxide, carbon monoxide in acid or alkaline CuCl, oxygen in alkaline pyrogallol (trihydroxybenzene—$C_6H_3(OH)_3$) and so on. Additional procedures can also be adopted for gas analysis, including titration, measurement of spectra (infra-red or ultra-violet), or CHROMATOGRAPHY.

gas laws the rules that relate to the pressure, temperature and volume of an ideal gas, allowing useful information about a gas to be gained by calculation instead of by experimentation. The laws are termed Boyle's law, Charles' law, and the pressure law. The pressure law states that, when a gas is kept in a constant volume, the pressure of that gas will be directly proportional to the temperature. All three laws can be combined in an equation known as the universal gas equation, which allows gases to be compared under different temperatures and pressures, i.e. $pV = nRT$, where p, V and T relate to pressure, volume and temperature respectively, n is the quantity of gas under investigation, and R is the universal molar gas constant, which has the value of $8.314 \, JK^{-1}mol^{-1}$.

gastric juice *see* **gastrin**

gastrin a hormone secreted in the stomach which stimulates the gastric glands of the stomach to produce gastric juice. The presence of food in the stomach is the initial stimulus, and the gastrin controls the digestive process. Gastric juice is a mixture of hydrochloric acid, certain salts and some ENZYMES, e.g. PEPSIN, that catalyse the breakdown of protein.

Gastropoda the largest class within the phylum Mollusca, with around 40,000 species. Although most are marine, there are freshwater varieties and some have adapted to land (garden snails, slugs, for example). The class also includes whelks, limpets and conches. Typical features include a large muscular foot used for locomotion (hence the name gastropod), a distinct head with eyes, tentacles, a rasping RADULA, and a coiled/twisted shell. The shell does vary in shape from the typical coil to the flatter limpet. Slugs have lost their shell and do not have the gills of their aquatic counterparts, the mantle cavity functioning as a crude lung.

gastrula the stage in animal embryonic development that involves the production of the GERM LAYERS. It follows formation of the BLASTULA. At this point the ball of the blastula becomes a multi-layered embryo with some cells moving to an inner location, forming a cup-shape with a cavity.

gelatin a pale coloured protein produced by boiling COLLAGEN with water and evaporating the resulting solution. It is water soluble, swelling when water is added and on cooling forms a gel from a solution in hot water. It is used extensively for preparing cultures in microbiology and also in pharmaceutical preparations and the food industry.

gel filtration *or* **gel permeation chromotography** molecules of different sizes can be separated by passage of a solution through a gel in a column. Molecules above a certain size cannot enter the pores of the gel, while smaller molecules move through the gel structure by diffusion. Those molecules excluded move quickly down the column. Smaller molecules move slowly through the gel, and at different rates depending upon molecular size and shape. The components of a mixture are thus separated in order of decreasing size or molecular weight. The technique is used to separate proteins.

gene the chemically complex unit of heredity, found at a specific location on a CHROMOSOME, that is responsible for the transmission of information from one generation to the next. Each gene contributes to a particular characteristic of the organism, and gene size varies according to the characteristic that it codes for. For example, the gene that codes for the HORMONE called insulin, consists of 1700 BASE PAIRS on a DNA molecule.

gene cloning a method of GENETIC ENGINEERING whereby specific genes are extracted from host DNA and introduced into the cell of another host by means of a plasmid VECTOR. All the descendants of the genetically transformed host cell will produce a copy of the gene. The transformed gene is thus said to have been cloned (*see also* CLONE and PLASMID).

gene flow the transfer of genes between populations via the GAMETES. Gene flow enhances variation in a population as it can lead to a change in the frequency of ALLELES present within that population. This in turn is a factor that contributes to EVOLUTION, as the alleles affect the characteristics of an organism. Therefore, gene flow can be advantageous, as it can help an organism inherit new characteristics that may be beneficial to its survival.

gene probe a technique used to detect a GENE, which employs BASE PAIRING between the gene and a sequence on another nucleic acid molecule, e.g. DNA or RNA. The single-stranded nucleic acid is called the *probe*, and it is labelled with a radioactive isotope so that it can be detected subsequently and the required gene identified and, if necessary, grown in culture. It is used in GENETIC ENGINEERING.

gene sequencing the determination of NUCLEOTIDE sequences of DNA molecules through the use of RESTRICTION ENZYMES. The enzymes are used to cut long DNA molecules into smaller fragments for identification. Each fragment is labelled with a radioactive phosphate, or fluorescent dye, for subsequent identification. The fragments undergo specific reactions and then ELECTROPHORESIS, permitting elucidation of the base sequence of the fragments. This is the Maxam-Gilbert method. Another technique, the Sanger method, involves synthesizing DNA strands (*in vitro*) which are complementary to strands being sequenced. The synthesis can be stopped at chosen points, at any of the four bases (nucleotides). As with the first method, labelled fragments can be separated and identified. Most gene sequencing is now automated.

gene therapy the use of GENETIC ENGINEERING to correct defective GENES. It is particularly suited to diseases due to the deficiency of a single ENZYME. Normal genes can be introduced into those cells that continue to reproduce throughout life so that the disorder is corrected and the normal gene replicated.

genetic adaptation *see* **adaptation**.

genetic code the code whereby instructions for protein synthesis are encoded in DNA. Because there are just four nucleotides, base triplets are the smallest units that can provide codes for all 20 AMINO ACIDS. It has been confirmed that the transfer of information from gene to protein is based upon a triplet code, i.e. genetic instructions can be written as a series of three-nucleotide terms and those are called *codons*.

genetic engineering the branch of biology that involves the artificial modification of an organism's genetic make-up. The term covers a wide range of techniques, including selective plant and animal breeding, but it is especially associated with two particular techniques:

(1) The transfer of DNA from one organism to a

different organism in which it would not normally occur. For example, the gene that codes for the human hormone, insulin, has been successfully incorporated into the GENOME of bacterial cells, and the bacteria produce insulin.

(2) Recombination of DNA between different species in the hope of producing an entirely new species. For instance, cells of the potato and tomato plants, which have had their cell walls removed, have been successfully cultured and made to fuse together using a variety of experimental procedures. Such cells can grow successfully and develop into a new species of plant that has been called the pomato. Although crossing the species barrier is an important breakthrough in the field of genetic engineering, there are strict governmental regulations regarding the release of such species into the environment since the consequences cannot be predicted.

genetic recombination the exchange of genetic material during meiosis, with the effect that the resultant GAMETES have gene combinations that are not present in either parent. This rearrangement of genes allows for variability in a species, and in each generation an almost infinite variety of new combinations of ALLELES of different genes are created. Such novel combinations of genes can confer enormous benefits to an organism when conditions change. For example, only a tiny number of a population of locusts have specific combinations of genes that enable them to survive potent pesticides. When such insects reproduce, they produce resistant populations—a major problem in the world of agriculture.

genetics the study of heredity and variation in biology. The classical aspects of the subject were expounded by Mendel (*see* MENDEL'S LAWS OF GENETICS), but recent work has branched out into POPULATION GENETICS. In addition, research in microbiology and biochemistry has led to a greater understanding of GENES, REPLICATION and GENETIC ENGINEERING.

genome the total genetic information stored in the CHROMOSOMES of an organism. The number of chromosomes is characteristic of that particular species of organism. For instance, a man has 23 pairs of HOMOLOGOUS CHROMOSOMES (containing approximately 50,000 genes), domestic dogs have 39 pairs, and domestic cats have 19 pairs. In each case, one pair of chromosomes constitute the SEX CHROMOSOMES, and the remaining pairs are the AUTOSOMES.

genotype the specific versions of the GENES in an individual's genetic make-up. For instance, there are three possible genotypes for the human albino gene, and it has two allelic forms, dominant A and recessive a. Thus, the three possible genotypes are:
(1) AA (homozygous dominant)
(2) aa (homozygous recessive)
(3) Aa (heterozygous).

genus a taxonomic division that includes a number of species with clear similarities and itself forms part of a family (*see also* SPECIES).

geochronology the study of time on a geologic scale through the use of *absolute* and *relative age-dating* methods. Relative dating deals with the study and use of FOSSILS and sediment to put rock successions or sequence in order. Absolute methods provide an actual age for a rock, using radioactive elements, which, because their decay rate (HALF-LIFE) is known, enables the necessary calculations to be made (*compare* RADIOCARBON DATING).

geology the scientific study of planet earth. This includes geochemistry, petrology, mineralogy, geophysics, palaeontology, stratigraphy, physical and economic geology.

geotropism a growth movement, exhibited by plants in response to the force exerted by GRAVITY. Plant roots are termed positively geotropic since they grow downwards, whereas plant shoots generally grow upwards (towards sunlight) thus displaying negative geotropism.

germ cell gamete-producing cells, i.e. part of the germinal epithelium. In mammals the germ cells are contained in the germinal epithelium of the ovaries and testes.

germination the start of growth in a dormant structure, e.g. a seed or spore. Various factors can break seed dormancy, such as specific temperatures, exposure to light, or rupture of the seed coat, all of which depend on the species from which the seed is derived.

germ layers layers of cells at the GASTRULA stage of an animal embryo. Each layer, ECTODERM (outer), ENDODERM (inner) and where present the MESODERM, give rise to the various organs of the body.

gestation period the period from conception to birth in mammals, which is characteristic of the species concerned. For instance, dogs have a gestation period that on average is 63 days, whereas that of the blue whale is 11 months.

giant axons *or* **giant fibre** large-diameter nerve axons that occur in many invertebrates and some vertebrates, including earthworms, crustaceans and the squid where fibres may reach 1 millimetre. The axons permit rapid conduction of impulses, enhanced by fewer SYNAPSES. The reason for these fibres is to permit the rapid transmission of impulses required, for example, when responding to a threat and immediate escape is imperative.

gibberellin (*see also* HORMONES) plant GROWTH SUBSTANCES, obtained originally from a fungus, which have marked effects in certain plants. In the original case some rice seedlings in Asia grew much taller than the rest and eventually the fungus was shown to secrete a chemical. Specifically, they boost cell elongation and are involved in fruit growth and seed germination. There are more than 70 gibberellins based on a common molecule, and often there is a synergistic interaction with AUXIN. Gibberellins also promote growth of buds in spring by blocking another hormone (abscisic acid) which inhibits plant growth.

gill outgrowths of the body surface that through their large surface area and enhanced blood supply are specialized for gaseous exchange. Gills may be external or internal, simple or complex. In some echinoderms gills appear as small lumps on the surface and in some segmented worms (ANNELIDs) each segment possesses a pair of flaps which are gills. Other animals possess gills in one location on the body, occurring as feathered, frilled and/or convoluted surfaces providing a large area for respiration. This system is found in molluscs, crustaceans, some amphibians and, of course, fish, where water is pumped in through the mouth and out of the GILL SLITS and the gills are protected by a flap called the OPERCULUM. Gills must operate very efficiently to obtain oxygen from water, as its concentration is less than that in air.

gill slit a long, slit-like opening in aquatic vertebrates, from the pharynx to the exterior. In some animals they are used in filter feeding whereas in fish they contain the gills. Gills are a characteristic embryonic development in chordates (*see* CHORDATA).

gizzard a part of the ALIMENTARY CANAL in some animals where food is ground up before the main process of digestion. The gizzard of birds contains grit or small stones that, with the muscular walls, assist in breaking up food.

glaciation a term meaning ice age, with all its effects, processes and products. The most recent is associated with the Pleistocene (*see* APPENDIX 5), but the rock record indicates older glaciations from the Precambrian and Permo-Carboniferous, and other periods in geological history.

gland an organ, group of cells or sometimes a single cell, occurring in both animals and plants that secretes a particular substance or substances. Animals have endocrine (*see* ENDOCRINE SYSTEM) and exocrine glands, the former secreting into the blood. Exocrine glands secrete onto an epithelial surface via a duct.

Plant glands secrete into a cell, cavity, or duct, or to the outside, and one example is the digestive gland of certain CARNIVOROUS PLANTS.

glial cells supporting cells in the CENTRAL NERVOUS SYSTEM, of which there are several types. *Astrocytes* surround capillaries in the brain and assist in the control of the chemical environment of the central nervous system. *Oligodendrocytes* form the MYELIN SHEATHS around AXONS and in the PERIPHERAL NERVOUS

SYTEM the myelin sheath is formed from SCHWANN CELLS. These supporting cells are present in ten to fifty times the number of neurons in the nervous system.

globulins globular PROTEINS occurring widely in plants (as a reserve protein in seeds) and in milk, blood and eggs. Globulins in blood serum are of four types: α_1, α_2, β and γ. The alpha and beta types are carrier proteins (like haemoglobin), and gamma-globulins include the IMMUNOGLOBINS that are invoked in immune responses.

glomerulus a set of blood capillaries that is surrounded by the cup-shaped end of a KIDNEY tubule. The cup end is called the Bowman's capsule after the British physician who discovered it. Small molecules (solutes, e.g. water and waste) are filtered through the capillaries under pressure from the blood into the Bowman's capsule and down the nephron.

glucagon an important HORMONE in the maintenance of the body's blood sugar levels. Glucagon works antagonistically with INSULIN, stimulating an increased supply of blood sugar through the hydrolysis of GLYCOGEN and it prompts the production of glucose in the liver from amino acids and fatty acids. Glucagon is secreted by the ISLETS OF LANGERHANS in the PANCREAS in response to a low blood sugar level. The balance to be maintained is 90 mg/ml and disfunction in the glucagon/insulin mechanism leads to DIABETES.

glucocorticoids one of two main types of CORTICOS-TEROIDS in humans, its primary action being on GLUCOSE metabolism and resistance to extreme physiological conditions such as starvation, prolonged cold. Glucocorticoids facilitate glucose synthesis from (non-carbohydrate) substrates to render more glucose available as fuel in a stressful situation, and promote GLYCOGEN deposition in the liver. It is administered medically to suppress certain aspects of the IMMUNE SYSTEM.

gluconeogenesis a major metabolic pathway, occuring predominantly in the liver, that synthesizes glucose from non-carbohydrate precursors in conditions of starvation. Glucose is required by red blood cells and is the primary energy source of the brain. However, the glucose reserves present in body fluids are sufficient to meet the body's needs for only about one day. Therefore, gluconeogenesis is very important during longer periods of starvation or during periods of intense muscle exercise. There are three major non-carbohydrate classes that serve as raw materials for gluconeogenesis:
(1) GLYCEROL—this is derived from fat hydrolysis.
(2) AMINO ACIDS—this is derived from protein degradation during starvation and from proteins in the diet.
(3) Lactate (lactic acid)—this is formed by actively contracting muscle when there is an insufficient supply of oxygen. (It is also produced by red blood cells).

glucose the most abundant naturally occurring sugar, which has the general formula $C_6H_{12}O_6$. Glucose is distributed widely in plants and animals and is an important primary energy source, although it is usually converted into polysaccharide carbohydrates, which serve as long-term energy sources. The storage polymers of plants and animals are starch and GLYCOGEN respectively. Other polysaccharides of glucose include CHITIN and CELLULOSE, which have a structural role and also provide strength.

glutathione an antioxidant PEPTIDE occurring widely in animals, plants and micro-organisms. It is made up of the amino acids glutamic acid, cysteine and glycine, and in its reduced form reacts with harmful oxidizing agents. This process is important in maintaining the proper activity of proteins, haemoglobin, etc.

gluten a protein mixture (glutenin and gliadin) that occurs in ENDOSPERM of wheat grain and is responsible for the 'firmness' of risen dough in bread-making. The amino acid composition of the proteins varies but glutamic acid and proline are major components. Some people suffer from *coeliac disease* which is an allergy to gluten (a sensitivity of the intestine lining) resulting in a dietary deficiency caused by poor absorption of food components.

glyceride a fatty acid ESTER formed from GLYCEROL. Any or all of the three hydroxyl groups of glycerol can be replaced giving rise to mono-, di- or triglycerides. Triglycerides are important as a major component of fats and oils providing a food energy store in living organisms. They are also used as cooking fats, margarines etc. If one of the hydroxyls is replaced with a phosphate group it forms a phosphoglyceride (a PHOSPHOLIPID) and the addition of a sugar forms a GLYCOLIPID.

glycerol a viscous, sweet-smelling alcohol, which has the chemical formula $HOCH_2CH(OH)CH_2OH$. Glucose is widely distributed in plants and animals as it is a component of stored fats. During metabolism, stored fats break down to form the original reactants, glycerol and FATTY ACIDS, while a large amount of energy is released. Glycerol is used commercially to manufacture a wide range of products, including explosives, resins, toilet preparations and foodstuffs.

glycogen often called animal starch, a polySACCHARIDE of GLUCOSE units which occurs in animal cells (especially the muscle and liver) and acts as a store of energy released upon hydrolysis. Glycogen is also found in some fungi.

glycolipid a lipid containing mono- or oligosaccharides and found in plasma membranes. Composition and complexity vary and they may be involved in cell-surface recognition.

glycolysis a major metabolic process, occurring in the CYTOPLASM of virtually all living cells, where the breakdown of glucose into simple molecules generates energy in the form of ATP. Each 6-carbon glucose molecule is converted into two 3-carbon pyruvate molecules ($CH_2COCOOH$) in a sequence of ten reactions, giving a net gain of two ATP molecules. The glycolytic pathway is regulated by several ENZYMEs. Although the reactions converting glucose to pyruvate are very similar in all living organisms, the fate of pyruvate is variable. In AEROBIC organisms, pyruvate enters the MITOCHONDRIA, where it is completely oxidized to CO_2 and H_2O in a process known as Krebs cycle (or CITRIC ACID CYCLE). This cycle, together with glycolysis, liberates 38 molecules of ATP per glucose molecule. However, if there is an insufficient supply of oxygen, e.g. in an actively contracting muscle, FERMENTATION occurs and pyruvate is converted into lactic acid, liberating only 2 ATP molecules per glucose molecule. In some

ANAEROBIC organisms, such as YEAST, pyruvate is converted into the alcohol ETHANOL during fermentation, again yielding only 2 ATP molecules. If a cell requires energy, or certain intermediates of the pathway are required for the synthesis of new cellular components, glycolysis proceeds, provided that glucose levels in the blood are high. However, when blood-glucose levels are low, e.g. during starvation, glycolysis is inhibited and instead GLUCONEOGENESIS occurs. Glycolysis and gluconoegenesis are reciprocally regulated so that when one process is relatively inactive, the other is highly active.

glycoprotein a protein which has a sugar residue linked to it by covalent bonding (*see* GLYCOSYLATION). Acidic glycoproteins (proteoglycans) occur in the matrix around cells of animal tissues (especially connective tissues). Cell surface glycoproteins are usually short, complex oligosaccharides bound to a membrane protein (*see also* GLYCOSYLATION).

glycosylation the process by which a carbohydrate is added to an organic compound, for example a protein. Glycosylation may occur in the ENDOPLASMIC RETICULUM or the GOLGI APPARATUS of cells and plays an important role in regulating protein activity.

Gmelin test a test used to determine the presence of BILE pigments that results in the formation of coloured oxidation products.

gnotobiotic a term describing germ-free conditions, applied commonly in experimental laboratories.

Golgi apparatus a system of ORGANELLES within the cells of organisms, comprising stacks of flattened sacs that act as the assembly point for the modification, sorting and packaging of large molecules; proteins, for example, undergo GLYCOSYLATION here.

Numerous small membrane-bound vesicles surround the Golgi apparatus, and these are thought to transport the modified macromolecules from the Golgi apparatus to the different compartments of the cell. It is named after the Italian physician Camillo Golgi (1844–1926) who discovered its existence.

gonadotrophin *or* **gonadotrophic hormone** a group of HORMONES secreted by the anterior PITUITARY GLAND. Both sexes produce follicle-stimulating hormone (FSH) and LUTEINIZING HORMONE (or interstitial cell-stimulating hormone in males) which stimulate reproductive activity in the gonads (ovaries and testes). Chorionic gonadotrophin is produced by the placenta, maintaining the CORPUS LUTEUM.

gonads the reproductive organs of animals, which produce the GAMETES and certain HORMONES. The male and female organs are known as the testes and the ovaries respectively.

Gondwanaland the massive, hypothetical, continent in the southern hemisphere that gave rise to parts of the present Africa, South America, India, Australia, New Zealand and Antarctica. Their connection at one time is postulated as a reason for the occurrence of widely separated but similar groups of plants and animals.

Graafian follicle a fluid-filled cavity in a mammalian OVARY that protects the developing egg cell. After the ovum is released the follicle forms the CORPUS LUTEUM.

graft when a piece of living tissue is joined to another tissue in the same or a different host. Subsequent growth fuses the tissues together. Grafting is used a great deal in horticulture to propagate certain bushes and fruit trees. A bud or shoot (scion) from the variety chosen is grafted on to the rootstock of another plant. The graft retains its characteristics and the usual processes of photosynthesis and water/mineral transport occur.

In humans and animals, grafts are used to replace damaged or faulty organs or tissues. A *homograft* involves taking from one individual and giving to another, i.e. transplantation, a process that is surgically complex and open to possible rejection of the "graft" by the host immune system.

gram (g) the basic unit of mass. There are approximately 28g in an ounce, and precisely 1000g in a kilogram.

Gram's stain a stain used to differentiate between bacteria, based upon differences in the structure of their cell walls. The Danish bacteriologist H. C. J. Gram first described the technique in 1884 and it involves, firstly, staining bacteria on a microscope slide with a violet dye and iodine. The slide is then rinsed in ethanol (to decolourize) and a second (red) stain is added. The structure of the bacteria cell walls differ in that the simpler, Gram-positive varieties, have a plasma membrane covered by a cell wall of peptidoglycan (polysaccharides and polypeptides linked together to form a network) whereas Gram-negative bacteria have less peptidoglycan located in a layer between two plasma membranes. The violet dye is retained in Gram-positive varieties but is washed out of Gram-negative forms, which then retain the red dye.

granum structures (*plural* **grana**) a structure found in the chloroplasts of plants and consisting of stacks of disc-like vesicles, resembling piles of coins, called thylakoids. Grana carry CHLOROPHYLL and contain enzymes involved in the light reactions of PHOTOSYNTHESIS. The thylakoids are the sites for the reactions.

graph a diagram that represents the relationship between two or more quantities, using dots, lines, bars or curves.

graptolite an extinct marine organism which secreted CHITINOUS tubes attached to branches, forming small (several cm) colonies. They first appeared in the Middle Cambrian and form useful INDEX FOSSILS for the Ordovician and Silurian.

gravity the attractive force that the Earth exerts on any body that has mass, tending to cause the body to accelerate towards it. Other planets also exert a force of gravity, but the force is different from that exerted by the Earth since it depends on the planet's mass and diameter. The true WEIGHT of any object on Earth is really equal to the object's MASS (m) multiplied by the acceleration due to gravity (g), which is 9.8 ms^{-2}. Therefore, although weight and mass are often used synonymously, they are different for scientific purposes. For example, a man with a mass of 80kg will weigh 784 newtons (N) on Earth, but on the Moon he would weigh only 130N since the force of gravity on the Moon is only $\frac{1}{6}$th of that on Earth. However, his mass is still 80kg and remains constant throughout the universe.

greenhouse effect the phenomenon whereby the Earth's surface is warmed by solar radiation. Most of the solar radiation from the Sun is absorbed by the Earth's surface, which in turn re-emits it as infrared radiation. However, this radiation becomes trapped in the Earth's atmosphere by carbon dioxide (CO_2), water vapour and OZONE, as well as by clouds, and is re-radiated back to Earth, causing a rise in global temperature. The concentration of CO_2 in the atmosphere is rising steadily because of mankind's activities (e.g. deforestation and the burning of FOSSIL FUELS), and it is estimated that it will cause the global temperature to rise 1.5–4.5°C in the next fifty years. Such a rise in temperature would be enough to melt a significant amount of polar and other ice, causing the sea level to rise by perhaps as much as a few metres. This could have disastrous consequences for coastal areas, in particular major port cities like New York.

grey matter a component of the CENTRAL NERVOUS SYSTEM of vertebrates, specifically the spinal cord and brain. It is brown-grey in colour and is the co-ordination point between the nerves of the central nervous system. It is composed of nerve cell bodies, DENDRITES, SYNAPSES, GLIAL CELLS and blood vessels.

groundwater water contained in the voids within rocks. It usually excludes water moving between the surface and the WATER TABLE (*vadose* water) but may be meteoric, of atmospheric origin, or juvenile water (original water from within the Earth).

growth hormone (somatotrophin or FH) a HORMONE of the PITUITARY GLAND that controls protein synthesis and growth of long bones in legs and arms. It is transported by a GLOBULIN protein in plasma. In humans, low levels produce dwarfism and overproduction gigantism.

growth rings the rings seen in a cross-section of a woody stem, for example, a tree. Each ring represents one season's growth (or XYLEM). Pale wood is produced in the early part of the season, with slower late growth producing a dense, darker wood (often called spring and autumn wood respectively). The ring width varies with the conditions, being dependent upon light, water, nutrients, and temperature. In addition to allowing an estimation of a tree's age to be made from the number of rings, each ring is different, and comparison of dead and living trees provides a dating method (and information about the climatic history) for periods within the past eight thousand years.

growth substance a botanical term referring to plant hormones whether natural or synthetic. Organic chemicals are produced in certain parts of the plant, often in just a small patch of cells. Included are AUXIN, ETHENE and GIBBERELLINS.

guanine ($C_5H_5N_5O$) a nitrogenous base component of the nucleic acids, DNA and RNA. It has a PURINE structure with a pair of fused HETEROCYCLIC rings, that contain nitrogen in addition to carbon. In both DNA and RNA, guanine always BASE PAIRS with CYTOSINE, which has a pyrimidine structure. Guanine is also present in many other biologically important molecules.

gum arabic *or* **acacia gum** a white, water-soluble powder, which in natural form is obtained from some varieties of acacia trees. It is a complex polysaccharide (*see* SACCHARIDE) and is used widely in pharmacy as an emulsifier (*see* EMULSION), and as an adhesive. It is also used in the food industry as an emulsifier and inhibits sugar crystallization.

Gymnospermae *or* **Gymnophyta** a division of the SPERMATOPHYTA, containing the conifers and similar species (e.g. cycads and the extinct seed ferns). The class is found as fossils back to the Devonian. The

Gymnospermae

term gymnosperm means "naked seed" and, unlike the ANGIOSPERMS, the OVULE and seed are not enclosed in a CARPEL. Fertilization is carried out by wind-borne pollen, and in general this group is not as advanced as the angiosperms.

H

habitat the place where a plant or animal normally lives, specified by particular features, e.g. rivers, ponds, sea shore.

habituation a type of learning in which, if a stimulus is continuously or repeatedly applied, an animal's respose to it declines.

haem *see* **myoglobin**

haemocyanin any one of a number of respiratory PROTEINS, containing copper, that are found in solution in the haemolymph (blood) of many MOLLUSCS and ARTHROPODS. Oxygen is reversibly bound by two copper atoms contained in haemocyanin. It changes from the colourless deoxygenated form (Cu I) to the blue coloured oxygenated form (Cu II) and functions in a similar way to HAEMOGLOBIN.

haemoglobin an iron-containing red pigment, which is found within the red blood cells (or ERYTHROCYTES) of vertebrates and which is responsible for the transport of oxygen around the body. In actively metabolizing tissue, e.g. the muscles, haemoglobin exchanges oxygen for carbon dioxide (CO_2), which is then carried in the blood back to the heart and pumped to the lungs, where the haemoglobin loses the CO_2 and regains oxygen.

haemolysis the rupture of red blood cells and release of HAEMOGLOBIN, which may be brought about, for example, by the presence of bacterial toxins or poisons or because of certain allergic reactions. It may result in anaemia.

haemophilia a genetic disorder affecting the blood, in which the lack of a vital BLOOD CLOTTING factor causes abnormally delayed clotting. Haemophilia is exhibited almost exclusively by males, who receive the defective gene from their mothers. A haemophilic female can only arise if a haemophilic male marries a female carrying the gene (extremely rare). There is no known cure for haemophilia, and, when injured, haemophiliacs must rely on blood transfusion to replace the blood loss, which is considerably greater than that lost by a normal individual.

haemopoiesis refers to the formation of whole BLOOD in vertebrates, including both cells and plasma (*see also* MYELOID TISSUE).

haemostasis *see* **blood clotting**.

hair in plants, a single-celled or multicellular thread-like outgrowth from the EPIDERMIS, e.g. the trichome of blue-green algae. In mammals, a thread-like dead structure consisting of many keratinized cells (*see* KERATIN) that arises from the epidermis of the skin. The hair root lies beneath the surface of the skin and is contained within a HAIR FOLLICLE. The hair bulb at the base of the follicle produces the hair cells. The colour of hair is determined by the amount and presence of the pigment MELANIN. In cold conditions an animal's hair is raised by erector pili muscles and traps a layer of air that has a warming effect.

hair follicle a cylindrical pit of small diameter in mammalian SKIN, which contains the root of a hair.

It extends down through the EPIDERMIS and DERMIS of the skin with its base in the SUBCUTANEOUS tissue and is lined with epidermal cells. Ducts of one or more SEBACEOUS GLANDS usually empty into the hair follicle.

half-life (t) the time taken for a radioactive ISOTOPE to lose exactly half of its RADIOACTIVITY. The half-life is constant for a particular isotope, varying from a fraction of a second to millions of years, and is best determined by using a Geiger-counter (Geiger tube). For instance, if an isotope has a half-life of one minute, then the radioactive count will fall by one half in one minute, by one quarter in two minutes, by one eighth in three minutes, and so on.

halophilic bacteria bacteria that can tolerate salt and live in the surface layers of the sea. They are instrumental in various biochemical cycles including the nitrogen and carbon cycles.

halophyte a plant that can tolerate a high level of salt in the soil. Such conditions occur in salt marshes, tidal river estuaries (and on motorway verges and central reservations!) and a typical species is rice grass (*Spartina*).

haltere present in the two-winged flies (DIPTERA), it is one of a modified pair of hind wings that act like gyroscopes and are concerned with maintaining flight stability. A haltere consists of a lobe at the base arising from the thorax, elongating into a stalk and an end knob, and it can only move in the vertical plane. There are numerous sense organs (campaniform sensilla) on the basal lobe that are sensitive to movements and forces generated during flight, and transmit the information to the thoracic GANGLION. In this way, adjustments of the wings can be made.

The halteres beat at the same rate as the fore wings but out of phase with them, and it is their vibration that causes the characteristic buzzing sound produced by flies during flight.

hammer *see* **malleus**.

haploid this term describes a cell nucleus or an organism that possesses only half the normal number of CHROMOSOMES, i.e. a single set of unpaired chromosomes. This is characteristic of the GAMETEs and is important at fertilization as it ensures the DIPLOID chromosome number is restored. For example, in a man there are 23 pairs of chromosomes per somatic cell, which is the diploid number, but the gametes possess 23 single chromosomes, which is the haploid number.

hard water water that does not readily form a lather with SOAP. This is due to dissolved compounds of calcium, magnesium and iron. Use of soap produces a scum, which is the result of a reaction between the FATTY ACIDS of the soap and the metal ions. The scum is made up of SALTS that, when removed, render the water soft. There are two types of hardness. Temporary hardness is created by water passing over carbonate rocks (e.g., limestone or chalk), producing hydrogen carbonates of the metals, which dissolve before the water reaches the mains supply. Boiling the water decomposes the hydrogen carbonates into carbonates (producing kettle fur), and the water becomes soft. Permanent hardness in water is due to metal sulphates, which can be removed by the addition of sodium carbonate. Zeolites will remove both types of hardness.

Hardy-Weinberg ratio a law that states that in a large, randomly breeding population, the genetic and allelic frequencies will remain constant from generation to generation. For example, a particular GENE in a population may have a number of ALLELES, one of which is dominant, but this does not necessarily mean that it occurs at a higher frequency than the recessive alleles. If the gene has two alleles, B (dominant) and b (recessive), present at frequencies x and y respectively, then the proportion of the genotypic frequencies would be:

$$BB \quad Bb \quad bb$$
$$x^2 \quad 2xy \quad y^2 = 1.0$$

There are certain conditions of stability that must be met for such a genetic equilibrium to occur:

(1) The population must be large, so that allelic frequencies could not be altered by chance alone.

(2) There must be no mutation, or it must occur in equilibrium.

(3) There must be no immigration or emigration, which would alter the genetic frequencies in question.

(4) Mating and reproductive success must be completely random with respect to GENOTYPE. If all conditions of the Hardy-Weinberg law are met, EVOLUTION could not occur as allelic frequencies would not change. However, evolution does occur because the conditions are never entirely met. For example, the condition of random mating is probably never met in any real population since an organism's genotype almost always influences its choice of a mate and the physical efficiency and frequency of its mating. The condition regarding mutation is probably never met either, since mutations are always occurring, resulting in a slow shift in the allelic frequencies in the population, with the

more mutable alleles tending to become less frequent.

Haversian system an anatomical feature of compact BONE, consisting of a narrow tube, the Haversian Canal, which is surrounded by concentric rings of bone (lamellae). The Haversian canals contain nerves and blood vessels and connect with one another throughout the bone.

heart a hollow, muscular organ that acts as a pump to circulate blood throughout the body. The heart lies in the middle of the chest cavity between the two lungs. It is divided into four chambers, known as the right and left ATRIA, and the right and the left VENTRICLES. In normal persons there is no communication between the right side and the left side of the heart, thus the two sides act as independent pumps, which are connected in series. Starting from the left ventricle, the flow of blood is as follows:

(1) Left ventricle contracts and oxygenated blood is pushed into the AORTA under pressure.

(2) Aorta divides into numerous ARTERIES to supply blood to all parts of the body.

(3) Deoxygenated blood returning from the body is carried by small VEINS, which eventually join up to form two large veins, called the superior VENA CAVA and the inferior vena cava.

(4) These two large veins empty into the right atrium.

(5) The blood passes from the right atrium to the right ventricle via a VALVE.

(6) The right ventricle contracts, pushing blood under pressure into the PULMONARY ARTERY.

(7) The pulmonary artery branches into two, carrying blood to both the right and left lungs.

(8) Within the lungs, gas exchange occurs—carbon dioxide is expelled, and the blood is oxygenated (*see* HAEMOGLOBIN).

(9) The blood flows from the left atrium via a valve into the left ventricle.

heartwood the central wood of a tree trunk or branch consisting of dead XYLEM tissue that does not conduct water or nutrients but provides mechanical support. The heartwood often contains resins, oils and gums that make it darker in colour and harder than the surrounding SAPWOOD.

heat ENERGY produced by molecular agitation.

heat capacity (C) the quantity of heat required by a substance or material to raise its temperature by one degree KELVIN (or one Celsius). Thus, heat capacity is measured in joules per Kelvin (JK^{-1}) or joules per Celsius (JC^{-1}). The molar heat capacity of a substance is the heat required that will raise the temperature of one MOLE of the substance by one degree, and the specific heat capacity is the heat capacity per kilogram ($JK^{-1}kg^{-1}$) or gram ($JK^{-1}g^{-1}$).

heat exhaustion a physical state experienced by warm-blooded animals whereby the body's normal cooling processes fail to operate as a result of increasing environmental temperature. Instead, the body's metabolic rate increases, raising the body temperature higher, which in turn raises the metabolic rate even higher, and so on. The symptoms of heat exhaustion are cramp and dizziness, and death ensues when the body temperature reaches about 42°C, which is the upper lethal temperature for the average human being.

heat of combustion the amount of heat generated when one MOLE of a substance is burned in oxygen.

heat of dissociation the amount of heat required to dissociate (*see* DISSOCIATION) one MOLE of a compound.

heat of formation the heat required or given out when one MOLE of a substance is formed from its elements (at one atmosphere and usually 298K).

heat of reaction the amount of heat absorbed or given out for each MOLE of the reactants. If the reaction is exothermic (i.e. heat given out), the convention is to specify the quantity as a negative figure (in kilojoules).

heat of solution the amount of heat absorbed or given out when one MOLE of a substance is dissolved in water.

hela cell a particular cell variety, discovered in a woman with cervical carcinoma (a form of CANCER) in 1951. These transformed cells are immortal and are used in laboratories worldwide for research purposes.

helminth a common term used to describe parasitic worms (e.g. Platyhelminthes and Nematoda) *see* TURBELLARIA, TREMATODA and CESTODA.

Hemiptera a large order of insects (*see* INSECTA) comprising the true bugs, e.g. leaf hoppers, scale insects, bed bugs, cicadas, aphids and water boatmen. The typical body shape is a flattened oval with two pairs of wings folded flat on the abdomen in the resting insect. The anterior wings are either hard throughout—the Homoptera, or only at the bases with membranous tips—the Hemiptera. Mouth parts are tubes modified for piercing and sucking and some are herbivorous, others carnivorous and yet others parasitic. Some are very significant agricultural pests transmitting plant diseases through feeding on sap.

hepatic portal system a system present in cephalochordates (marine animals such as *Amphioxus*), but typical of vertebrates, it refers to the system of capillaries and veins that carry the products of digestion absorbed into the blood from the intestine directly to the liver. The main VEIN involved is the hepatic portal vein.

hepatitis a serious disease of the liver, of which there are two types:

(1) Hepatitis A (infectious hepatitis) caused by the hepatitis A virus.

(2) Hepatitis B (serum hepatitis) caused by the hepatitis B virus.

Both diseases share the same symptoms of fever, nausea and JAUNDICE, but they are transmitted by different routes. Hepatitis A is spread by the oral-faecal route and occurs in people who have poor sanitation and personal hygiene. The virus can be transmitted from person to person in contaminated food or drinking water. Most people exposed to the disease can be protected by PASSIVE immunization, which involves the administration of purified ANTIBODIES from a previously infected individual who has recovered. Hepatitis B is spread through blood products, contaminated syringes and instruments. Susceptible groups include those who require blood or blood products, e.g. haemophiliacs (although any donated blood is normally screened for hepatitis). A significant percentage of hepatitis B sufferers develop cancer, and the virus is thought to be a contributory factor.

herbaceous "herb-like," describing a plant with little woody tissue in which the aerial growth dies back at the end of each season. Either the whole plant dies, as in annuals, or parts of the plant survive beneath the ground to resume growth the following year.

herbivore any animal that feeds on vegetation (*see* CARNIVORE, OMNIVORE).

hermaphrodite (bisexual) in flowering plants, it refers to the commonest situation where flowers contain both STAMENS and CARPELS (*see* ANGIOSPERMAE). In animals it describes an individual that has both male and female sex organs and is common among some vertebrates. In certain animals, such as parasitic worms (*see* CESTODA), self-fertilization takes place. In others, such as the earthworm (*see* ANNELIDA), an exchange of sperm takes place between two individuals.

herpes one of a group of disease-causing VIRUSES that are responsible for infections in man and other vertebrates. They are complicated viruses, containing DNA, which tend to cause recurring infections. Examples are the Epstein-Barr (EB) virus, which causes glandular fever and probably Burkitt's lymphona (a cancer); *Cyto megalovirus* (*see* CONGENITAL) and *Herpes simplex*, the cause of cold sores. *Herpes varicella/zoster* is the causal agent of chicken pox and shingles.

Hess's law a law that states that, irrespective of the intermediate stages, the overall heat change in a chemical reaction depends only on the beginning and final states.

heterocyclic compounds organic compounds forming a ring structure with the additional elements, e.g. oxygen, hydrogen, nitrogen and sulphur.

heterodont of dentition, referring to animals that have more than one type of tooth (MOLARS, PREMOLARS,

CANINES, INCISORS) each of which performs a different function. This is the commonest type of dentition found in mammals (*see also* HOMODONT).

heterotrophism a type of nutrition describing an organism that must take in organic material, either plant or animal, in order to provide for its own bodily needs. Food is ingested and digested, and provides organic molecules and energy upon which the organism's existence depends. All animals are heterotrophs (*see* AUTOTROPHISM).

heterozygote an organism having two different ALLELES of the GENE in question in all somatic cells. For instance, if gene B has two allelic forms, B and b, then the heterozygote will contain both alleles, i.e. Bb, at the appropriate location on a pair of HOMOLOGOUS CHROMOSOMES. Heterozygotes can thus produce two kinds of GAMETES, B and b. One allele of a heterozygote is usually dominant, and the other is usually recessive. The dominant allele is the one that is expressed phenotypically, because it masks the expression of the recessive allele. Dominant alleles are usually denoted by capital letters, while recessive alleles are denoted by lower case letters.

hexose a MONOSACCHARIDE sugar, the molecules of which contain six carbon atoms. Examples are GALACTOSE, GLUCOSE and FRUCTOSE and they are important sources of energy (*see* SACCHARIDE).

hibernation an adaptation seen in some animals, in which a sleep-like state occurs that enables an animal to survive the cold winter months when food is scarce. It is accompanied by some marked physiological changes, notably a drop in body temperature to about 1°C above the surrounding temperature, and a slowing of the pulse and metabolic rate to about 1 per cent of normal. In this state an animal's energy requirements are reduced and are provided for by stored body fat. It is common in bears, bats, rodents, hedgehogs and also fish, amphibians and reptiles.

Hill reaction the light-dependent stage of PHOTOSYN-THESIS, in which illuminated CHLOROPLASTS initiate the photochemical splitting of water. This produces hydrogen atoms (two per water molecule), which are used to reduce carbon dioxide with the formation of carbohydrate in the dark stage of photosynthesis. The light stage also produces ATP, which provides the energy required for carbohydrate synthesis.

hind brain *or* **rhombencephalon** one of the three sections of the embryonic vertebrate BRAIN, lying behind the MIDBRAIN and FOREBRAIN, from which develops the CEREBELLUM, PONS and MEDULLA OBLONGATA.

hip girdle *see* **pelvic girdle**.

hippocampus this lies over each of the two lateral VENTRICLES (cavities) in the vertebrate brain, and consists of two ridges or tracts of nervous material running back from the olfactory lobe to the posterior part of the CEREBRUM. It is very well developed in the advanced mammals, such as cetaceans (*see* CETACEA) and primates, and appears to be involved in the generation of emotional responses as well as having a role in the memory pathway.

histamine an AMINE released in the body during allergic reactions, and in injured tissues. Release causes dilation of blood vessels, causing a fall in blood pressure.

histochemistry the study of the distribution of chemical molecules within tissues by means of chemical analysis. Such techniques as light and electron MICROSCOPY, auto-radiography, chromatography and STAINING are employed.

histocompatibility the extent to which tissue cells from one animal will be accepted by the IMMUNE SYSTEM of another animal, which is medically very important. The immune system possesses the ability to recognize foreign material, whether a tissue GRAFT or invading micro-organisms, because the invaders have HISTOCOMPATIBILITY ANTIGENS on their surface. Each individual organism has a set of histocompatibility antigens unique to itself so that the immune system can recognize and reject "non-self" (*see* ANTIGEN).

histocompatibility antigens GLYCOPROTEINS occurring on or near the surface of tissue cells. In humans and mice (which have been studied extensively) these antigens are determined by a complex gene cluster called the MAJOR HISTOCOMPATIBILITY COMPLEX (MHC). Two distinct types of histocompatibility antigens are present. Class I occur on the surface of most body cells and help CYTOTOXIC T-CELLS recognize virus-infected cells which then destroy them. Class II occur only on certain inducer cells (LYMPHOCYTES) of the immune system such as B-CELLS, and MACROPHAGES where they act as receptors for foreign antigens. The inducer cells present the foreign antigen to helper T-cells, leading, by a series of activities within the immune system, to rejection of the foreign material (*see also* HLA SYSTEM).

histogram a graph that represents the relationship between two variables using parallel bars, but it differs from a bar chart in that the frequency is not represented by the bar height, but by the bar area.

histology the study of the tissues, tissue structure and organs of living organisms, in the main through microscopic techniques.

histolysis breakdown of a cell or tissue.

histone one of a number of very important water soluble PROTEINS containing the positively charged (basic) AMINO ACIDS arginine, lysine and histidine. They are involved in the packaging of the DNA of plant and animal CHROMOSOMES and, together with the DNA, form chromatin. Chromatin forms most of the structure of the chromosome. Histones are absent from bacterial cells and vertebrate sperm cells.

HIV (*abbreviation for* human immune deficiency virus) the retrovirus thought to be the cause of AIDS.

HLA system (*abbreviation for* human leucocyte antigen system) one group of HISTOCOMPATIBILITY ANTIGENs that are coded for by a series of four gene loci (A, B, C and D on chromosome 6 in human beings). These are Class I antigens and are of major medical importance in determining the degree of acceptance or rejection of GRAFTed organs or tissues. Histocompatible individuals possess identical HLA types, and transplants are more successful if the number of differences between the HLA systems of donor and recipient can be kept to a minimum (*see* MAJOR HISTOCOMPATIBILITY COMPLEX).

Holocene (Recent) the most recent geological epoch of the Quaternary period, which started after the final glaciation of the Pleistocene (about 10,000 years ago) up to and including the present day. Hence sometimes called the post-glacial epoch, and considered by some scientists to be an interglacial interlude of the Pleistocene to be followed by another glaciation.

holoenzyme an enzyme molecule and its cofactor which together form a complex that is catalytically active. The enzyme (protein) component of the complex is inactive on its own and is known as an APOENZYME (*see* COFACTOR and ENZYME).

holophyte an organism that has plant-like nutrition; effectively the same term as photoautotroph (*see* AUTOTROPHISM).

holotype *or* **type specimen** the individual organism used to describe and name a SPECIES (*see* NEOTYPE). Usually this is or has been done within the system of binomial nomenclature, which is the classical method of TAXONOMY.

holozoic feeding like animals, i.e. organisms that are heterotrophic (*see* HETEROTROPHISM).

homeostasis the various physiological control mechanisms that operate within an organism to maintain the internal environment at a constant state. For example, homeostasis operates to keep the body temperature of humans within a small, crucial temperature range, independent of the temperature of the external environment, as our metabolic processes would not function in any other temperature range.

homeothermy *see* **homoiothermy**.

hominid a member of the primate family Hominidae, which includes modern man as well as fossil hominids that may have evolved around 3.5 million years ago, all belonging to the genus *Homo*. The characteristic feature is bipedalism, i.e. walking upright, and also adaptations in the shape of the skull and teeth. Development of a culture is also apparent (*see* CROMAGNON MAN).

homo primate genus that includes modern man

Homo sapiens and various other extinct species. There is much controversy over the classification of fossil types but *H. erectus* and *Homo habilis* are usually placed in this genus.

homodont animals in which the teeth are all of the same type and typical of all the vertebrates except mammals (*see* HETERODONT).

homoiothermy *or* **homeothermy** *or* **warmbloodedness** the ability of an animal to sustain a constant body temperature, independent of the temperature of its environment, that involves various metabolic processes. It is typical of birds and mammals and evidence sugggests that some of the DINOSAURS were homoiothermic (*see* HYPOTHALAMUS and POIKILOTHERMY).

homologous pertaining to organs or structures that have evolved from a common ancestor, regardless of their present-day function. For example, the pentadactyl limb is the ancestral form of the quadruped forelimb, and from it evolved the human arm, the fin of cetaceans, and the wings of birds. These structures are therefore said to be homologous. Similarities in homologous structures are best seen in early embryonic development and imply relationships between organisms living today.

homologous chromosomes chromosomes that are identical in their genetic LOCI but can have individual allelic forms (*see* ALLELE) that are not necessarily the same. In DIPLOID organisms, a pair of homologous chromosomes exists in all SOMATIC CELLS, each member of the pair having come from a different parent. During MITOSIS in somatic cells, homologous chromosomes do not associate with each other in any way. However, during MEIOSIS in the formation of the GAMETES, homologous chromosomes join together to form a pair, and exchange of genetic material may occur before the homologous pair separates into two new cells that produce gametes. Therefore, the gametes have only a single set of chromosomes so that at fertilization the diploid number is restored.

homologous series chemical compounds that are related by having the same functional group(s) but formulae that differ by a specific group of atoms. For instance, the ALKENES form an homologous series in which each successive member has an addtional CH_2 group, i.e.:

Alkene series	Molecular formula
Ethene	C_2H_4
Propene	C_3H_6
Butene	C_4H_8

homozygote an organism that has two identical ALLELES of the GENE in question in all SOMATIC CELLS. For instance, if gene B has two allelic forms, B and b, the homozygote will contain only one allelic type, i.e. either BB or bb, at the appropriate location on a pair of HOMOLOGOUS CHROMOSOMES. Homozygotes can thus only produce one kind of gamete, B or b, and as such are capable of pure breeding. For example, the gene for albinism is RECESSIVE, and any individual that possesses this phenotypic trait will be homozygous for the gene. If two such individuals breed, the resultant offspring will all be albinos.

hormone an organic substance, secreted by living cells of plants and animals, that acts as a chemical messenger within the organism. Hormones act at specific sites, known as "target organs," regulating their activity and eliciting an appropriate response.

In animals, hormones are secreted from various ductless glands, which include the pancreas, thyroid and adrenal glands. This hormone-signalling system is collectively known as the ENDOCRINE SYSTEM. These glands secrete hormones directly into the bloodstream, usually in small amounts, where they circulate until they are picked up by appropriate receptors present on the cell membranes of the target organs. These receptors recognize the particular hormone and bind to it, initiating a response. Hormones also play an important part in the role of plant and seed growth and are found in root tips, buds, and other areas of rapid development. For example, gibberellins are a class of plant hormones involved in processes such as initiating responses to light and temperature, the formation of fruit and flowers, and the promotion of seed elongation. Hormone action is constantly regulated by elaborate FEEDBACK MECHANISMS, both within and between cells and organs, that regulate their secretion and breakdown.

host describes an organism that inadvertently supports another organism within its own body or attached to it. The host organism is damaged to a greater or lesser extent by this arrangement and it describes the host-parasite relationship. A host may be primary (definitive) containing the sexually mature parasite or secondary (intermediate) containing an immature larval stage (*see* PARASITE).

humerus the main long bone of the forelimb of tetrapod vertebrates. It articulates with the shoulder blade (SCAPULA) of the pectoral girdle and with the ULNA and RADIUS at the elbow.

humidity the amount of water vapour in the earth's

atmosphere. The actual mass of water vapour per unit volume of air is known as the absolute humidity and is usually given in kilograms per cubic metre (kgm⁻³). However, it is useful to use relative humidity, which is the ratio, as a percentage, of the mass of water vapour per unit volume of air to the mass of water vapour per unit volume of saturated air at the same temperature.

humus material that makes up the organic component of soil, being formed from decayed plant and animal remains. It has a characteristic dark colour and its chemical composition varies according to the amount and type of material present. It has colloidal properties (*see* COLLOID) and hence holds water which is then made available to growing plants. Similarly, being a colloid it helps to prevent the loss of minerals from the soil (by leaching) and hence is very important in determining soil fertility. Humus may be more acidic (mor) as in the soil of coniferous forests, or alkaline (mull) found in the soil of deciduous woodlands and mixed grasslands. Humus contains numerous micro-organisms and invertebrate species, and its presence is of obvious economic importance in the cultivation of food crops.

Huntington's chorea *see* **lethal gene**.

hybrid the offspring of a mating between two parents that are not genetically similar, i.e. they differ in at least one characteristic. It can be used in a more restricted sense in genetics but generally refers to the offspring of a mating between two different species, which, in animals, is usually sterile. The most familiar example is the mule, the result of a mating between a female horse (mare) and a male donkey (stallion).

hybrid vigour describes the situation where the HYBRID progeny possess greater vigour than either of the parent types. Hybrid crop varieties often have a better yield than either of the parent plants. The mule is a more robust and longer-lived animal than either the horse or donkey, although it is sterile.

hydrocarbon an organic compound that contains carbon and hydrogen only. There are many different hydrocarbon compounds, the most common being the ALKANES, ALKENES, and ALKYNES.

hydrochloric acid an aqueous solution of hydrogen chloride gas, producing a colourless, fuming, corrosive liquid. It will react with metals to form chlorides, liberating hydrogen. It is made by the ELECTROLYSIS of brine producing hydrogen and chlorine, which are combined, or by the reaction of sulphuric acid with sodium chloride. It has many uses in industry.

hydrocortisone *see* **cortisol**.

hydrogenation an important industrial reaction where gaseous hydrogen is the vehicle for adding hydrogen to a substance. The reaction usually proceeds in the presence of a CATALYST and often at elevated pressure. This reaction is utilised in the petroleum refining and petrochemicals industry, the hydrogenation of coal to produce HYDROCARBONS and the hydrogenation of fats and oils.

hydrogen bond electrostatic attraction between an electronegative atom—oxygen, nitrogen or fluorine—in one molecule and a hydrogen atom attached to some other electro-negative atom in another molecule. It is a relatively weak non-covalent bond, but is of enormous biological importance. Hydrogen bonds are a significant component of

PROTEIN structure and for BASE PAIRING between NU-
CLEIC ACID strands in DNA chains. They account for
the properties of water and for its high boiling point.

hydrolysis the term used to describe a chemical
reaction where the action of water causes the decomposition of another compound and the water
itself is decomposed. In salt hydrolysis, the salt
dissolves in water, producing a solution that may be
neutral, acidic or basic, depending on the relative
strengths of the ACID and BASE of the salt. For
example, a solution of potassium chloride (KCl)
would be neutral, since potassium forms a strong
base and chlorine forms a strong acid. In comparison, ammonium chloride (NH_4Cl) gives an acidic
solution, since ammonium forms a weak base but
chlorine forms a strong acid.

hydrophyte any plant that inhabits extremely wet
conditions, such as water-logged soil, or actually
grows in water. These plants are specialized, having little supporting or vascular tissue and very
little or no root system. The leaves are also highly
adapted with little or no CUTICLE. Examples are
pondweeds and water lilies (*see* MESOPHYTE, XERO-
PHYTE).

hydroponics a system of commercial cultivation of
some crop plants. It is developed from water culture
methods used in the laboratory where plants are
selectively deprived of certain minerals and the
effects on growth are assessed. The roots of the
plants being cultivated are immersed not in soil but
a solution containing the correct balance of essential mineral nutrients.

hydrosphere the water that exists on or near to the
Earth's surface. The main components are water

(H_2O), sodium chloride (NaCl) and magnesium chloride ($MgCl_2$). By mass, the major elements are oxygen (almost 86 per cent), hydrogen (10.7 per cent), chlorine (2 per cent) and sodium (1 per cent). Magnesium is the only other element present in significant quantities.

Hydrozoa a class of marine invertebrates belonging to the phylum CNIDARIA, containing several different orders and including corals (millepore corals), hydroids (e.g. *Hydra*) and siphonophores (e.g. *Velella*, *Physalia*). An alternation of generations is common (*see* ASEXUAL REPRODUCTION) between a sessile POLYP phase and a free-swimming medusoid phase. However, in different species either phase may be reduced or absent. Most polyps are colonial and some MEDUSAe also form floating colonies. The gonads are ectodermal (*compare* SCYPHOZOA), and most are marine forms, but some, e.g. *Hydra*, live in fresh water.

Hymenoptera an important order of the class INSECTA containing the bees, wasps, ants, ichneumon flies and sawflies. They usually have a "wasp waist"—the abdomen is first constricted and then enlarged posteriorly. The first abdominal segment is fused to the thorax and the forewings are larger than the hind wings. The hind wings and forewings interlock by a series of tiny hooks present on the former's leading edges. However, wings are absent in some species. The mouthparts are variously adapted for biting, sucking or lapping: an ovipositor is used for egg laying and may be adapted for piercing, stinging or sawing and is often long and looped forward. Development typically involves the production of pupae often within cocoons, and PARTHENOGENESIS is common. POLYMORPHIC and social species occur, notably the ants and some bees and wasps. The honey bees have always been an important species to man.

hyperplasia the abnormal increase in the size of an organ or tissue due to a proliferation of cells by cell division. An example is the growth of a tumour (*see* HYPERTROPHY).

hypertonic a solution that exerts a higher osmotic pressure than another solution due to the presence of a greater concentration of solute (*see* HYPOTONIC, OSMOSIS).

hypertrophy an increase in the size of an organ or tissue brought about by an enlargement in the dimensions, rather than the number, of its cells. It may occur as a result of an increased workload in an organ, due to disease or other malfunction.

hypha (*plural* **hyphae**) a delicate, hair-like, tubular filament present in fungi. Hyphae may be packed tightly in an interwoven mass (pseudoparenchyma) or loosely intertwined (mycelium). They may be branched and may or may not contain inner cross walls (SEPTA). The cell walls may be of CHITIN or CELLULOSE with CYTOPLASM in which GLYCOGEN and oil globules often occur. There is usually a central VACUOLE, and growth is by an increase in length occurring at the tip of the hypha. ENZYMES are secreted near the tips of hyphae, which act on the fungal substrate to provide nutrients (*see* PARASITE, SAPROPHYTE).

hypodermis lying immediately below the EPIDERMIS in the leaves of certain plants, this is the outermost layer of cells in the CORTEX. The cells are often modified and strengthened to provide mechanical support and protection, and may store water or nutrients.

hypogeal refers to seed germination when the cotyledons (seed leaves) remain underground. Examples are pea, broad bean and oak. It also refers to such fruiting bodies as peanuts and truffles that develop beneath the soil surface.

hypophysis *see* **pituitary gland**

hypothalamus a part of the vertebrate brain that develops from the third ventricle of the FOREBRAIN. It lies beneath the CEREBRUM and THALAMUS on the ventral surface and regulates a wide variety of functions. In higher vertebrates these include feeding, sleeping, thirst, body temperature and blood pH, and the hypothalamus also contains a centre concerned with emotions such as aggression. These are brought about by means of the AUTONOMIC NERVOUS SYSTEM, which is controlled by the hypothalamus, and the NEUROENDOCRINE SYSTEM. The hypothalamus produces releasing factors and hormones that are transported to the PITUITARY GLAND, which, in its turn, secretes hormones that act on other glands in the body. Hence the hypothalamus indirectly controls various endocrine functions. The cells involved are specialized NEURONS that receive nerve IMPULSES from other nerve cells but respond by releasing hormones. Two sets of these neurosecretory cells are found within the hypothalamus: one set produces the releasing factors (hormones) that act on the anterior pituitary; the other produces the hormones of the posterior pituitary itself, received from the hypothalamus via the blood stream.

hypotonic refers to a solution that has a lower concentration of solute than another solution and therefore has a lower osmotic pressure (*see* HYPERTONIC, OSMOSIS).

I

ice age the spreading of glacial ice over areas that are normally ice-free. Most is known about the Pleistocene ice age, but the geological record contains evidence of earlier events although the evidence is less clear.

identical twins the offspring that develop from a single fertilized ovum that splits into two very early during development, producing two separate individuals. Identical twins (also known as monozygotic twins) have precisely the same genetic constitution and are always of the same sex.

Ig *abbreviation for* IMMUNOGLOBIN.

ileum the part of the SMALL INTESTINE after the JEJUNUM (although anatomically indistinct from it) and before the LARGE INTESTINE. It is the region specialized in digestion and absorption of nutrients. The surface is covered with many projections called villi (*singular* VILLUS).

ilium the largest of the bones that form each half of the pelvic girdle. The others are the ISCHIUM and PUBIS. The ilium has a flattened wing-like part that fastens to the SACRUM by LIGAMENTS.

imago the sexually mature adult form in the life cycle of an insect.

immiscible the term that describes liquids that

cannot be mixed together. Such liquids tend to be polar and non-polar (e.g. water and ether respectively), and, when added together, form two separate layers, with the less dense liquid forming the upper layer. Conversely, two polar liquids, or two non-polar liquids, will mix together and as such are termed miscible.

immune system the defence system within the bodies of vertebrates that evolved to afford protection against the pathogenic effects of invading micro-organisms and parasites. The immune system confers two types of immunity to an organism:

(1) Innate (or natural) immunity—this is present from birth and is non-specific, operating against almost any substance that threatens the body.

(2) Acquired immunity—this type of immunity is as a consequence of an encounter with a foreign substance, and it is specific against that foreign substance.

immunity *see* **immune system**

immunization rendering an individual immune (or less susceptible) by artificial means. Active immunization involves the introduction of treated bacteria, etc, to stimulate ANTIBODY production. Passive immunization is due to the injection of preformed antibodies, e.g. from an individual already immune to a particular ANTIGEN (*see also* IMMUNE SYSTEM).

immunoassay any technique for measuring the amount of a substance present that is reliant upon the sample binding to a particular ANTIBODY as would an ANTIGEN. For example, a specific antibody is attached to a substrate and the sample added. Any antigens thus added will bind to the antibody. Then a second antibody, matching a different site on the antigen is added, this time with some sort of label. When this binds to the antibody, it can be identified and the concentration of the test (antigen) sample determined (*see also* ELISA).

immunoglobin (Ig) these are groups of proteins, collectively termed ANTIBODIES, that are produced by specialized cells of the blood, which are called B-CELLS and which can bind to specific ANTIGENS. B-cells are stimulated to divide in the presence of particular antigens, and the resultant daughter cells produce quantities of immunoglobins, which play an important role in the body's IMMUNE SYSTEM.

impermeable a term to describe a substance that does not allow gases or liquids to pass through.

implantation *or* **nidation** the process whereby a fertilized (mammalian) egg becomes attached to the UTERUS wall. Thereafter it continues to develop. The egg passes into the uterus after fertilization in the FALLOPIAN TUBE as a ball of cells—the blastocyst. The outer cells of the blastocyst create a depression in the uterus wall into which it sinks.

imprinting a particular form of learning in young animals. In the early stages of their development they learn to recognize, and become attached to, their mother. However, it has been proven that any moving object larger than themselves can become the focus of this imprinting, including man. The Austrian zoologist Konrad Lorenz (1903–89) first described the phenomenon and produced a well-known study with greylag geese. Lorenz divided a clutch of eggs, half staying with the mother and half being put in an incubator. On hatching, the young with the mother showed normal behaviour, but the geese hatching from the incubated eggs spent their

first hours with Lorenz. As a result, those geese
followed Lorenz and did not recognize their own
mother or other adult geese and as adults they
preferred human company.

Other examples of imprinting are known including the "chemical" imprinting of salmon, which
enables them to return to the river of their origin to
spawn. Song development in birds can also be considered as imprinting, as extensive experimentation has shown.

Imprinting is further defined by a critical period
within which the learning occurs. Not only is imprinting irreversible, but if there is nothing on
which to imprint during the critical period, then the
animal in question will not imprint on anything else
afterwards.

impulse *or* **nerve impulse** the transmission of
information along a nerve fibre through the crea-
tion of an ACTION POTENTIAL.

inbreeding sexual reproduction between closely
related partners. The extreme example is the self-
fertilization of plants. Inbreeding affects the genotypic frequencies predicted by the HARDY-WEINBERG
RATIO and in general less variation is exhibited than
with a normal (or "outbreeding") population, and
ultimately harmful traits may be manifested (called
inbreeding depression, the opposite of HYBRID VIG-
OUR). In humans, mental and other defects are more
frequent with marriages of close relatives (e.g. cous-
ins).

incisor a sharp tooth in most mammals that is edged
like a chisel and is used for biting food. In some
animals it is used for gnawing or nipping and
modifications include the tusks of elephants.

incompatibility the rejection by the host, due to an immune response, of a GRAFT or blood transfusion. In botany, it is when the scion will not graft on to the rootstock, or pollen from a flower does not fertilize other flowers on the same or similar plants. This latter case promotes cross-fertilization.

incus the middle of the three EAR OSSICLES in the middle ear of mammals. Also called the anvil.

indefinite inflorescence *see* **racemose inflorescence**.

independent assortment a process suggested by the Austrian monk Gregor Mendel (1822–84) to explain the random distribution of different gene pairs that allows all possible combinations to appear in equal frequency. Independent assortment is also known as MENDEL'S second LAW OF GENETICS.

index fossil a fossil that is selected for a particular feature and that occurs in a particular zone. The zone usually takes the name of the fossil. An index fossil should ideally have a wide distribution in area but not in time. Over the fossil-bearing record, index fossils include TRILOBITES (in the Cambrian) and GRAPTOLITES (Ordovician and Silurian).

indicator a chemical substance, usually a large complex organic molecule, that is used to detect the presence of other chemicals. Indicators are usually weak ACIDS, the un-ionized form (often written as HA), having a different colour from the ionized form (H and A), because of the negative ion A. The degree of ionization, and thus colour change, depends on the pH of the solution under investigation. In solution, the indicator partially dissociates, i.e. HA —> H and A. Most useful indicators give a distinct colour change over a small pH range, usually about

two units. Commonly used indicators include phen-
ophthalein and methyl orange.

indicator species certain plant and animal species
are particularly dependent upon their immediate
environment and its overall quality. Their sensitivity to a specific factor in the environment may be
such that they will be affected by the presence, or an
excess of, this factor. For example, some LICHENs are
indicator species because they are sensitive to the
pollutant sulphur dioxide. Thus absence of a lichen
from an area where it would normally be expected to
grow could indicate sulphur dioxide pollution.

indigenous the term for a species that occurs naturally in a particular area, rather than being introduced by man.

indoleacetic acid (IAA) an AUXIN and the most
common plant GROWTH SUBSTANCE. It is produced in
shoot tips.

infection when a PATHOGEN enters a living organism
via a wound or through a mucous membrane (e.g.
lining the respiratory or alimentary tracts), and
through reproduction causes symptoms in the host.
The pathogen may be introduced by a carrier, an
insect vector or an infected individual. In animals
there is usually an incubation period after which
the symptoms appear. The body's IMMUNE SYSTEM
combats the infection, and antibiotic drugs can be
used against all but viral infections (*see also* IMMU-
NIZATION).

inflammation the body's defence reaction in the
area of tissue injury whether caused by infection or
physical damage, e.g. a cut. The initial response is
caused by the release of HISTAMINE from damaged
white cells. Histamine then causes the dilation of

small blood vessels (vasodilation) around the injury, a process enhanced by the release of other substances (prostaglandins), and clotting agents are also delivered to the area. The vasodilation and increased blood flow cause the typical redness and soreness of inflammations.

The next stage, which occurs within an hour of the injury, is the chemotactic (*see* CHEMOTAXIS) migration of PHAGOCYTES from the blood to the tissues. MACROPHAGES (a phagocytic cell) devour the pathogens, tissue cell remains and also remains of neutrophils, which are phagocytic cells that destroy microbes and themselves in the process. The dead cells often form pus at the infected site.

inflorescence the particular arrangement of flowers on a plant. There are several types of inflorescence divided into two groups. Racemose inflorescence is when the tip of the stem goes on producing new flowers on lateral branches. When the main stem ends in a flower with no further branching, it is called cymose inflorescence.

inhibition the effect of an inhibitor in reducing or restricting a biochemical reaction catalysed by an ENZYME. If the inhibitor molecule is similar to the substrate, it may bind to the enzyme's ACTIVE SITE or it may combine with the enzyme-substrate intermediate, thus removing it from the reaction. *Feedback inhibition* is due to products of the reaction and is a control factor in enzyme activity.

innate behaviour the behavioural make-up that occurs in all individuals of the same species and sex.

inner ear the part of the ear, within the temporal bone of the skull, that contains the apparatus for hearing and balance. Two chambers contain the

COCHLEA (hearing) and SEMI-CIRCULAR CANALS (balance).

innervation the supply of nerves to and from an organ.

innominate bone the bone formed from the fusion of the ILIUM, ISCHIUM and PUBIS, thus creating one half of the PELVIC GIRDLE in adult mammals, birds and reptiles.

inoculation the injection of mildly infective pathogen to stimulate an immunity, a technique superseded by IMMUNIZATION with non-infective agents.

Insecta a class of ARTHROPODS with well over three quarters of a million species represented on land, in the air and in freshwater. The class is divided into 26 orders and includes beetles (COLEOPTERA), flies (DIPTERA), bees and ants (HYMENOPTERA), butterflies (LEPIDOPTERA) and grasshoppers (ORTHOPTERA). The body of an insect is divided into head, thorax and abdomen. The thorax has three segments bearing legs and possibly wings. The head is characterized by antennae, COMPOUND EYES and simple eyes (*see* OCELLUS). The mouthparts are adapted to enable insects to feed on a variety of food sources. The abdomen has eleven segments.

The oldest fossil forms are from the DEVONIAN but the development of flight in the Carboniferous led to an enormous diversification. Another significant factor in their success is the waterproof CUTICLE. Internal organs are quite complex, and include an open circulatory system (haemocoel), excretory organs called Malpighian tubules and a system of branched CHITIN-lined tubes comprising the tracheal system that carries oxygen directly to the cells. The nervous system consists of a pair of nerve cords (with ganglia), which meet in the head with the ganglia of other anterior segments.

Reproduction is usually sexual although PARTHE-NOGENESIS does occur. Development is often through larval and pupal stages with moulting. Insects are important factors, whether for good or ill, in many areas. Many varieties are pests or disease carriers but equally insects pollinate crops and produce honey and silk.

Insectivora an order of placental mammals that are mainly small and nocturnal. Included are the moles, shrews and hedgehogs. They have long snouts with tactile hairs and tweezer-like INCISOR teeth. Their diet is mainly but not exclusively insects. The order is quite primitive and has changed little since their emergence in the CRETACEOUS, about 130 million years ago.

instar the stage between moults (ECDYSES) in the larval development of an insect. There are often a number of larval instars before the pupal stage.

instinct part of the behavioural patterns of animals, the other being learning linked with environmental factors. Instinct is a built-in behaviour pattern that does not depend greatly on learning for its expression. It is thus seen in individuals throughout a species, and often takes the form of a fixed action pattern or FAP. A fixed action pattern is highly stereotyped INNATE behaviour which, once started, is carried to completion. It is begun by some stimulus (or *releaser*) which the animal (innately) detects, and the innate behaviour produces some kind of activity.

The classic example is the male three-spined stickleback fish protecting his territory. The releaser is the red underside of the intruder and the fish will attack models only vaguely resembling fish providing they have a red underside, but will not attack a realistic stickleback model without the red colouration.

It is generally agreed that all behaviour is a mix of genetic factors, i.e. instinct, with learning and it also depends upon environmental conditions.

insulin a pancreatic HORMONE that initiates glucose uptake by body cells and thus controls glucose levels in the blood. Insulin functions by stimulating certain PROTEINS found on the surface of cells within the vertebrate body to take up glucose, which would otherwise be unable to enter cells, as it is very hydrophobic. *Diabetes mellitus* is a condition in which the blood contains excessively high glucose levels due to an under-production of insulin, and the excess glucose is excreted in the urine. This condition can prove fatal, but sufferers can be successfully treated by insulin therapy.

integument the outer layer of an animal that usually consists of an epidermal layer with dead keratinized cells (as in vertebrates, *see* SKIN) or a CUTICLE, as in ARTHROPODS. In botany, it refers to the outer covering of an OVULE, which eventually forms the seed coat.

intercellular a term meaning "between cells." It often refers to the matrix secreted by cells, as in connective tissue.

intercostal muscles muscles, located between the ribs and around the lungs, that are essential in breathing, in conjunction with the DIAPHRAGM (*see also* VENTILATION).

interferon a cytokine (a molecule secreted by a cell

that regulates other cells nearby) PROTEIN that leads
to the synthesis of antiviral proteins in a cell under
attack from viruses. First discovered in 1957, three
human interferons have been identified; α from
white blood cells, β from the fibroblasts of connective tissue and γ-interferons from lymphocytes (*see*
INTERLEUKINS). Interferons secreted by an infected
cell, before specific antibodies appear, go into neighbouring cells, prompting the production of other
proteins that restrict viral replication.

interleukins one of several cytokines (*see* INTER-
FERON) that act between LEUCOCYTES. There are four
interleukins involved in a variety of functions including the recognition of ANTIGENS, enhancing the
action of MACROPHAGES and the production of other
cytokines, e.g. interleukin-2 promotes the production of, amongst other substances, γ-INTERFERON.
Interleukin-2 is also used in the treatment of
melanoma.

interphase *see* **mitosis**.

intestine the part of the ALIMENTARY CANAL between
STOMACH and ANUS where final digestion and absorption of food matter occur in addition to the absorption of water and production of faeces.

intracellular referring to anything located within
the cell and usually, therefore, within the plasma
membrane.

intron a non-coding DNA segment (NUCLEOTIDE sequence) that interrupts the coding sequence of a
GENE, primarily in EUCARYOTES, forming 'split' genes.
The introns are transcribed (*see* TRANSCRIPTION) into
MESSENGER RNA. The precise role of introns is unclear
but it is thought they may function in a cell regulatory role, controlling gene activity. It seems likely

that they enable different types of cells to produce different proteins from a common gene.

inulin a POLYSACCHARIDE, a polymer of FRUCTOSE molecules, that forms a food reserve in the roots or tubers of numerous plants.

in vitro a term applied to experiments or techniques undertaken in the laboratory, where biological or biochemical processes are carried out "in glass."

in vivo biological and biochemical processes that occur in a living organism or cell.

involution the process, during gastrulation (*see* GASTRULA), when cells turn in from the outside to the inside of the gastrula. Also, the reduction in size of an organ whether returning to normal size after a particular event or due to falling usage (as with ageing).

ion an ATOM or MOLECULE with a positive charge due to electron loss (a CATION) or a negative charge due to electron gain (an ANION). The process of producing ions is known as ionization, and it can occur in a number of ways, including a molecule dissociating into ions when it is added to a solution, or the formation of ions by bombarding atoms with radiation.

ion exchange the exchange of IONS of like charge between a solution and a highly insoluble solid. The solid (ion-exchanger) consists of an open molecular structure containing active ions that exchange reversibly with other ions in the surrounding solution without any physical changes occurring in the material. For example, ion exchange is also used to soften hard water by removing calcium ions. The water is passed through a column containing an exchange resin that contains sodium ions, and these are exchanged for the calcium ions in the hard water, leaving the water calcium-free.

ionizing radiation radiation with sufficient energy to ionize the medium through which it passes. The radiation may be high energy electrons, alpha-particles, etc, or electromagnetic radiation such as gamma-rays or X-rays. Ionizing radiations are used in medical imaging, where the interaction between the radiation and the tissues provides information. Energy is transferred particularly to water molecules in tissues that liberate free RADICALS, which in turn may react with biological molecules to alter the structure of amino acids, proteins and enzymes. These changes may lead eventually to cell death, thus the use of ionizing radiation and the information gleaned must be balanced with the disadvantages.

iris muscular tissue in the EYE of vertebrates and some cephalopods that is pigmented and lies between the CORNEA and the LENS. Light enters through the pupil, and the amount of light is controlled by the reflex reaction of the muscles (circular and radial) reducing and enlarging (respectively) the diameter of the iris. The colour of the iris is due to differing amounts of the pigment melanin (blue—little; brown—more).

irritability the capacity of all organisms to respond to environmental changes, whether the mechanical response of a protozoan organism or the more complex reaction of a multi-cellular animal with sense organs.

ischium one of the three bones that comprise each half of the PELVIC GIRDLE. It is the most posterior and in primates supports the weight when sitting.

islets of Langerhans endocrine cells throughout the PANCREAS. Two hormones are secreted—GLUCAGON and INSULIN from the α- and β-cells respectively. The islets were discovered by Paul Langerhans, a German anatomist.

isoelectric point the pH of a solution, at which the overall charge on protein molecules is nearest to zero. This means there is minimum repulsion between molecules and thus the protein will precipitate most readily at the isoelectric point, i.e. it is in the least soluble state.

isolating mechanisms any factor that inhibits total interbreeding within a species and thus limits exchange of genes. If different groups of the same species are isolated geographically in different habitats, this may eventually lead to the formation of new species through the process of ADAPTIVE RADIATION.

isotonic (solutions) solutions with the same osmotic pressure (*see* OSMOSIS).

isotope an atom that differs from other atoms of the same element due to a different number of NEUTRONS within its nucleus. As isotopes still have the same number of PROTONS, their ATOMIC NUMBER is unchanged, but the varying number of neutrons affects their MASS NUMBER. Most elements exist naturally as a mixture of isotopes but can be separated due to the fact that they have slightly different physical properties. For laboratory purposes, a device called a mass spectrometer is used to separate the different isotopes.

J

jaundice a condition characterized by the unusual presence of bile pigment circulating in the blood. Jaundice is caused by the BILE produced in the liver passing into the circulation instead of the intestines because of some form of obstruction. The symptoms of jaundice include a yellowing of the skin and the whites of the eyes.

jejunum lying before the ILEUM and after the DUODE-NUM, it is a part of the mammalian SMALL INTESTINE. Its main function is the absorption of digested food products, and its lining has numerous projections (VILLI) that increase the surface area over which this can occur. The villi are longer in the jejunum than elsewhere in the small intestine.

joint the area where one (or more) bone meets another bone in the vertebrate skeleton, together with the surrounding tissues. These fall into three different categories depending upon the amount of movement that can occur at a particular joint:

(1) Immovable joints occur between the flattened bones of the CRANIUM (skull) and are marked by suture lines.

(2) Partly movable joints can be found between the vertebrae of the spinal column.

(3) Freely movable or synovial joints are those that

articulate the limb bones and these include two different types: hinge joints, found at the elbow and knee, allow movement in one plane only; ball and socket joints, found, for example, at the hip and shoulder, allow all round movement in more than one plane. LIGAMENTS bind synovial joints, and they are lined with a synovial membrane, which contains synovial fluid that bathes the joint and helps to protect it.

joule the unit of all ENERGY measurements. It is the mechanical equivalent of heat, and one joule (J) is equal to a force of one NEWTON moving one metre, i.e. $1J = 1Nm$. It is named after James Prescott Joule (1818–1889), a British physicist who investigated the relationship between mechanical, electrical and heat energy, and, from such investigations, proposed the first law of THERMODYNAMICS, the conservation of energy.

jugular (vein) one of major paired veins that occur in mammals and other vertebrates. Usually there is a pair of interior and exterior jugulars running from the head (and brain), which fuse on each side to become the common jugular veins. These run through the neck and join the subclavian veins at its base, eventually draining into the VENAE CAVAE and hence returning blood from the head to the heart.

Jurassic the second geological period of the Mesozoic era from about 213 million years ago to 144 million years ago. It followed on from the TRIASSIC and was succeeded by the CRETACEOUS, and it was named after the rocks of the Jura mountains between Switzerland and France. Many fossils are found in the limestones and clays laid down during the Jurassic. Plant types include cycads, ferns, conifers and rushes. Animals include AMMONITEs, bivalves (BIVALVIA), brachiopods (BRACHIOPODA), echinoids (ECHINODERMATA) and corals. The dominant vertebrates were reptiles but with a decline in the mammal-like types such as therapsids. In particular, icthyosaurs and pleisiosaurs were abundant, and the primitive bird *Archaeopteryx* first appeared, as did the flying reptiles, the pterosaurs.

juvenile hormone *or* **neotenin** a hormone found in insects and secreted by a pair of glands located near the brain, the *corpora allata*. At each moult, this hormone suppresses the development of adult features (METAMORPHOSIS) and maintains the larval stage. At the final moult, the corpora allata become inactive so that metamorphosis can proceed (*see* ECDYSIS).

K

karyotype the number, shapes and sizes of the chromosomes within the cells of an organism. Every organism has a karyotype that is characteristic of its own species, but different species have very different karyotypes. For example, all normal human females have 22 pairs of DIPLOID chromosomes with similar shape and size, but all female horses have 32 pairs of diploid chromosomes with their own unique shape and size.

keel *or* **carina** a bony plate projecting from the breastbone (sternum) of flying birds and bats that serves for the attachment of the powerful flight muscles. It is not present on the sternum of most flightless birds such as the ostrich (*see also* AVES).

kelp a large, brown, algal seaweed found anchored to the sea bed below low tide level. Kelp is a source of iodine and potash.

Kelvin scale the unit of temperature (K) based on the temperature scale devised by the British physicist Lord William Kelvin (1824–1907). The Kelvin scale has positive values only, with the lowest possible unit of 0K, which is equal to -273.15°C or -459.67°F.

keratin a fibrous, sulphur-rich protein consisting of coiled POLYPEPTIDE chains, which occur in hair, hooves, horn and feathers.

kerogen a fossil organic material, bituminous in nature, which is found in oil shales. As with all oil shales, the HYDROCARBONS are produced by destructive DISTILLATION of the shales.

ketenes ORGANIC compounds with the group >C=C=O of which the first member is ketene, $CH_2 = CO$. They are very unstable and polymerize easily. They react with water to produce ACIDS, and with ALCOHOLS to give ESTERS.

ketone an ORGANIC compound that contains a C=O (carbonyl) group within the compound as opposed to either end of the compound. There are many forms of ketones, and their physical and chemical properties differ due to the presence of alkyl groups ($-CH_3$) or aryl groups ($-C_6H_5$) within the ketone molecule. Ketones can be detected within the bodies of humans when fat stores are metabolized to provide energy if food intake is insufficient. If these accumulate within the blood, the undernourished person will experience headaches and nausea. The presence of ketones in urine is called ketonuria.

ketone body (*see also* KETONE) one of several compounds produced by the liver due to the metabolism of deposits of fat within the body. Examples are hydroxybutyrate and acetoacetate produced from coenzyme A. These substances normally provide an energy source for the body's peripheral tissues via a process called ketogenesis. However, in abnormal conditions, where the supply of carbohydrates is reduced (commonly as a result of starvation or in DIABETES), the level of ketone bodies in the blood rises. Ketone bodies may then be present in the urine giving off a characteristic "pear-drops" (acetone) odour in the condition known as ketosis.

kidney the main excretory organ of vertebrates, which is one of a pair in mammals, situated in the abdomen. The kidneys remove nitrogenous wastes from the blood and adjust the concentrations of various salts so functioning in osmoregulation. In mammals the waste fluid formed is known as urine and blood is supplied to the kidney from the renal artery and leaves via the renal vein. The urine that is formed collects in a central chamber within the kidney known as the renal pelvis. This leads in turn to the URETER which extends down into the bladder (one ureter from each kidney draining urine).

The kidney has two distinct functional regions, an outer cortex and an inner medulla. Its regulatory and excretory functions are carried out by means of numerous tubules called *nephrons*. A nephron comprises a renal tubule and its associated blood vessels. Capillaries transport water, urea, salts and other small molecules (known as filtrate) into a cup-shaped expansion of the tubule—the *Bowman's capsule*. The capillaries form a knot or ball called the GLOMERULUS from which filtrate passes through three distinct regions of the tubule: the proximal convoluted tubule, the loop of Henle and the distal convoluted tubule. The convoluted parts of the tubule have many bends whereas the loop of Henle is like a hairpin curved at the base but with straight descending and ascending portions. The distal convoluted tubule empties into the collecting duct, which receives filtrate from numerous other tubules. The filtrate undergoes final processing here and, as urine, is emptied into the renal pelvis. In most vertebrates all the parts of each nephron are contained within the cortex, and there are no loops of Henle. In mammals and birds, loops of Henle are present in all nephrons. In humans most are cortical nephrons with reduced loops confined mainly to the cortex. A small proportion (about 20 per cent) have longer loops of Henle that extend into the medulla (juxtamedullary nephrons). An AFFERENT ARTERIOLE, branching from the renal artery supplies blood to the nephron and itself divides many times to form the capillaries of the glomerulus. The capillaries converge again where they leave the glomerulus to form an EFFERENT arteriole that divides to surround the proximal and distal convoluted tubules. Additional capillaries (the vasa recta) extend to surround the loops of Henle. Interstitial fluid bathes the capillary network and tubules and there is a continuous passage of substances between the two in both directions (via the fluid).

The different parts of the nephron are associated with the various functions of osmoregulation and excretion and three processes are involved: filtration, secretion and reabsorption. The human kidney contains about 80 km of tubules numbering in the region of one million. The two kidneys daily receive between 1000 and 2000 litres of blood and process about 180 litres of filtrate; 1.5 litres of urine is produced and the rest is reabsorbed.

kilocalorie a unit of heat used to express the energy value of food. One kilocalorie is the heat needed to raise the temperature of one kilogramme of water by 1°C. It is estimated that the average person needs 3000 kilocalories per day, but this requirement will vary with the age, height, weight, sex and activity of the individual.

kilogram a unit of mass (kg) that is equal to the

international prototype made of platinum and iridium stored in the French town of Sèvres.

kinase an ENZYME that catalyses the phosphorylation of its substrate from ATP, hence it transfers a phosphate group from the high energy phosphate compound ATP to another molecule. The recipient molecule is itself often an enzyme, which is then activated and able to work.

kinesis the response of an organism to a particular stimulus in which the response is proportional to the intensity of the stimulation.

kinetic energy the energy possessed by a moving body by virtue of its mass (m) and velocity (v). The kinetic energy (Ek) of any moving body can be determined using the following equation:

$$Ek = \frac{1}{2}mv^2$$

(the energy is in joules if m is kg and v ms^{-1})

As the kinetic theory of matter states that all matter consists of moving particles, it holds that all particles must possess some amount of kinetic energy, which will increase or decrease with the surrounding temperature (*see* EVAPORATION).

kingdom the highest, most all-embracing category of taxonomy into which organisms are placed. It comprises numerous PHYLA. Originally there were two kingdoms, Plantae and Animalia, but now others are recognized (*see* CLASSIFICATION).

kinin in animals it is any one of a group of PEPTIDES that are found in the blood, and are concerned with the local regulation of blood flow. They are associated with INFLAMMATION and cause local increases in the permeability of tissue capillaries. An example is bradykinin, which is a hormone secreted by the submaxillary salivary gland under the control of the PARASYMPATHETIC NERVOUS SYSTEM, and is apparently implicated in allergic reactions. In plants, cytokinin is related chemically to ADENINE (a purine) and is one of a number of plant GROWTH SUBSTANCES that stimulate cell division in the presence of AUXIN.

Klinefelter's syndrome a condition in which human males have the abnormal GENOTYPE of XXY rather than normal XY. This produces recognizable characteristics within the affected male, such as the development of breasts and smaller testes (resulting in reduced fertility). Klinefelter's syndrome occurs in approximately one in a thousand male births and is caused by nondisjunction of sex chromosomes during MEIOSIS.

kneejerk reflex a complex neural pathway in humans in which a blow just beneath the kneecap (patella) results in a rapid extension of the leg. The knee jerk is a response by the central nervous system to stimulation of sensory NEURONS located at the front of the thigh. The response leads to the contraction of one muscle (quadriceps, at the front of the thigh) and the inhibition of the contraction of another muscle (biceps, at the back). As the quadriceps muscle is the extensor, the leg straightens but is prevented from bending as the biceps muscle, which is the flexor, is inhibited from contracting.

Köppen classification a system of climatic classification developed between 1910 and 1936 based upon annual and monthly means of temperature and precipitation, and the major types of vegetation. The system comprises three orders or levels, beginning with the overall climate, e.g. warm, temperate and rainy (class C), which can then be categorized further as, for example—winter, dry produces class Cw. The third level qualifies temperature, for example, a hot summer would produce a classification Cwa.

The major climates are: tropical rainy (A); dry (B); warm temperate rainy (C); cold snowy (D); and polar (E). (*See also* THORNTHWAITE CLIMATE CLASSIFICATION).

Korsakoff's syndrome *or* **Korsakoff's psychosis** a neurological disorder first described by a Russian neuropsychiatrist called Sergei Korsakoff (1854–1900). The condition is characterized by gross defects in memory for recent events, disorientation, and no appreciation of time. Patients with Korsakoff's syndrome are unaware that there is a problem and are liable to confabulate. Although it can result from lack of vitamins or lead and manganese poisoning, Korsakoff's syndrome most commonly occurs as a complication of chronic alcoholism. It is caused by a dietary deficiency of vitamin B1 (thiamine), which is needed for the conversion of carbohydrate to glucose. The syndrome has been invaluable in neuropsychology, as it has helped to identify the brain regions involved in the memory processes of recall and recognition.

Krebs cycle (*see also* **citric acid cycle**) a cyclical series of biochemical reactions, involving ENZYMES and found in the MITOCHONDRIA of cells, that completes the metabolic breakdown of GLUCOSE molecules generating carbon dioxide and ATP. It is also known as the citric acid cycle (or tricarboxylic acid cycle-TCA) and is of fundamental importance to the METABOLISM of all AEROBIC organisms. It is the second, vital, stage in cellular respiration. The Krebs cycle involves eight steps in which acetyl CoA, the common starting point (derived from the breakdown of "fuel," i.e. glucose and also fatty acids) is first converted to citric acid (citrate). This is then reconverted to oxalo-acetate resulting in the production of two carbon dioxide molecules and ATP. For each molecule of acetyl CoA, 12 molecules of ATP are produced.

L

labelled compound a compound used in radioactive tracing, where an atom of the compound is replaced by a radioactive ISOTOPE, which can be followed through a biological or physical system by means of the RADIATION it emits.

lacrimal gland a gland present in the eyelid of the eyes of most of the higher vertebrates, which produces fluid (tears) that bathes the surface of the eye. The fluid is drained from the eye to the nose via the lacrimal duct.

lacteal a very small, blind-ended lymph vessel which is associated with a VILLUS that projects from the surface of the small intestine of vertebrates. It absorbs digested fats, forming milky fluid known as chyle, which is then transported via the thoracic duct to the bloodstream.

lactogenic hormone *see* **prolactin**.

lactose milk sugar found only in mammalian milk and manufactured by the MAMMARY GLANDS. It is composed of one GALACTOSE molecule bonded to one GLUCOSE molecule.

laevorotatory compound any substance that, when in crystal or solution form, has an optically active property that rotates the plane of polarized light to the left (anticlockwise).

Lamarkism an early theory of evolution devised by the French biologist, Jean Baptiste de Lamarck (1744–1829). He postulated that an organism acquired characteristics during its lifetime because of the conditions imposed upon it by its environment. Hence bodily characteristics that enhanced survival were strengthened and those that were not would ATROPHY due to disuse. These acquired characteristics were then thought to be passed on to the offspring. Hence a giraffe would have acquired its long legs and neck by stretching up to browse on the leaves of trees, according to Lamarkism. Although plausible superficially this theory has never been satisfactorily proved and was largely replaced by DARWINISM.

lamella (*plural* **lamellae**) one of a number of thin membranous layers of plate-like structures. (1) In vertebrates, it especially refers to a concentric ring of hard material which is found in BONE. (2) In fungi, such as mushrooms and toadstools, it refers to a spore-bearing gill on the underside of the cap. It also refers to one of folded paired membranes (thylakoids) within a plant CHLOROPLAST on which the photosynthetic pigments occur. In bryophytes it is one of a number of thin flaps of tissue forming a sheet of plates on the dorsal surface of the leaves or THALLUS.

lamina the thin, flattened portion of a leaf blade or petal of a plant. It also refers to the leaf-like part of the thallus in certain ALGAE such as kelp (*see also* LEAF).

large intestine a part of the vertebrate ALIMENTARY CANAL lying between the SMALL INTESTINE and the ANUS. It consists of three regions, the CAECUM, COLON and RECTUM.

larva (*plural* **larvae**) the juvenile form that occurs in many animals, especially invertebrates but also amphibians and fish, which is normally very different from the adult. A larva can usually feed and lead an independent existence but is generally incapable of reproduction. Commonly, it is the larval form which hatches from the egg and in aquatic animals, ciliated larvae often disperse away from the site of hatching in the PLANKTON. A larva develops into an adult by undergoing metamorphosis, which may be a gradual or fairly rapid process. Examples of larvae are the tadpoles of frogs, the trochophores of polychaetes and molluscs, and the caterpillars of butterflies.

larynx the anterior dilated portion of the TRACHEA (windpipe), which is found in TETRAPOD vertebrates at its junction with the PHARYNX. Plates of CARTILAGE occur within its walls and these are controlled by means of laryngeal muscles that open and close the GLOTTIS. In amphibians, reptiles and most mammals it contains the VOCAL CORDS, which are folds of membrane projecting from the walls. These vibrate to produce sounds, the pitch of which is altered by movement of the cartilage, under the control of the laryngeal muscles, which changes the tension of the vocal cords. The larynx is the "Adam's Apple" of humans.

Lassaigne's test a chemical test for the presence of nitrogen and also sulphur or halogens (e.g. bromine, chlorine, fluorine). The sample is heated with sodium in a test tube, quenched and ground, and on reaction with certain reagents a characteristic product or colour is produced.

latent heat the measurement of heat ENERGY involved when a substance changes state. While the change of state is occurring, the gas, liquid or solid will remain at constant temperature, independent of the quantity of heat applied to the substance (an increase in heat will just speed up the process). The specific latent heat of fusion is the heat needed to change one kilogram of a solid into its liquid state at the MELTING POINT for that solid. For example, the specific latent heat of fusion for pure, frozen water (ice) at 273K (0°C) is 334 kJkg^{-1}. The specific latent heat of vaporization is the heat needed to change one kilogram of the pure liquid to vapour at its boiling point. In the case of pure water again, at its boiling point of 373K (100°C), 2260kJkg^{-1} is the specific latent heat of vaporization needed to change water into steam.

lateral line the sensory system of fish and aquatic amphibians, which comprises sensory cells (neuromast organs) in a line along the body. Movements in the water affect the neuromasts thus creating nerve impulses.

latex a milky fluid that exudes from the cut surfaces of some trees and herbaceous plants such as the dandelion. Its composition is mixed containing mineral salts, proteins, sugars, oils and alkaloids and, on exposure to air it rapidly coagulates. The function of latex in the plant is not certain but it is thought to have a role in wound healing, protection and nutrition. Some species provide latex that is of commercial importance, notably the rubber tree.

Laurasia one of two continental masses produced by the rifting of PANGAEA. Laurasia was the northern area, which became North America, Greenland, Europe and Asia, and the splitting apart is thought to have occurred in the early Mesozoic (APPENDIX 5), about 240 million years ago.

law of conservation *see* **energy, thermodynamics**.

LD50 the amount of a toxic substance, which, when applied in a specific manner, will kill 50 per cent of a large number of individuals within a species.

LDL *abbreviation for* LIPOPROTEIN.

LDL-receptor *see* **cholesterol**.

L-dopa an intermediate substance in the synthesis of NORADRENALINE and ADRENALINE and an immediate precursor in the production of dopamine, a brain NEUROTRANSMITTER. It is found in the adrenal glands. The level of dopamine is reduced in the brain of sufferers from Parkinson's disease but cannot be administered directly as it does not reach the brain. However, L-dopa does reach the brain when administered either by injection or orally, and is of clinical importance in the treatment of Parkinsonism.

leaf a flattened structure, present in vascular plants and bryophytes, that is the main site for TRANSPIRATION and PHOTOSYNTHESIS. The leaves of bryophytes are simple, non-vascular structures that are not the same as those of higher plants. A leaf forms a group of tissues called the leaf buttress at the apex of the stem. It consists of a flat, broad, expanded portion, the LAMINA (leaf blade) divided by a central, stalk-like midrib which is the main vascular channel. This joins a leafbase, the site of attachment to the stem. Sometimes a leaf stalk (petiole) is present anterior to the leaf base. In higher plants each leaf has an axillary lateral bud and produces GROWTH SUBSTANCES, and leaves show a definite arrangement (*see* PHYLLOTAXY). Additionally there is a great variety of textures, shapes and sizes, vein arrangement and type of attachment to the stem. A leaf may have a continuous outline (simple leaf) or the lamina may be subdivided into lots of leaflets (compound leaf). Various kinds occur, e.g. COTYLEDONS (seed leaves), foliage leaves, which are the site of transpiration and photosynthesis, scale leaves, which lack CHLOROPHYLL (*see* BULB, CORM), bracts, which are modified leaves containing axillary flowers and floral leaves (carpels, stamens, petals and sepals, *see individual entries*) modified for reproduction.

learning the process by which, as a result of experience, an animal alters its response to a situation or set of circumstances. The animal's behaviour may alter permanently in a way that is usually of benefit to it, and learning allows for greater flexibility of response. It allows the animal to adapt to its environment and is commonly seen in the higher vertebrates. Various kinds of learning have been recognized and include habituation, insight learning and conditioning.

Le Chatelier's principle a statement relevant to chemical reactions, which predicts that if the conditions of a system in EQUILIBRIUM are changed, the system will attempt to reduce the enforced change by shifting equilibrium.

lecithin one of a number of PHOSPHOLIPIDS, a glycerophospholipid (phosphoglyceride) based on glycerol. Three hydroxyl groups (-OH) occur, two of which are ESTERified with two fatty acids and one with a phosphate group itself bound to the amino alcohol CHOLINE. Lecithins are the commonest phospholipids found in animals and also in higher plants but not usually in micro-organisms. They are an important constituent of cell membranes, a component of BILE and a SURFACTANT in vertebrate lungs.

legionnaires' disease an infectious disease caused by the bacterium *Legionella pneumophila*, which inhabits surface soil and water. It has also been traced in water used in air-conditioning cooling towers. The main source of infection is inhalation of air or water carrying the bacteria, and so far there is no evidence that it is transmitted from an infected to a non-infected individual. Legionnaires' disease is really a form of pneumonia, and thus its symptoms include shortness of breath, coughing, shivering and a rise in body temperature. Healthy individuals should fully recover from infection if treated with the antibiotic called erythromycin.

legume a dry fruit or pod that is characteristic of the Leguminosae, the family that includes the beans, peas, clover and gorse. It is formed from a single CARPEL which, when mature, splits into two halves and opens to liberate the seeds.

lens a biconvex, transparent and usually crystalline structure found in the EYE of many animals. In vertebrates it is composed of numerous fibrous layers of the protein crystallin, and its main function is to focus light rays onto the RETINA. It is attached by suspensory ligaments to the ciliary body and divides the eye into two cavities, one anteriorly between the lens and the CORNEA and a much larger one posteriorly within the eyeball itself. Radial muscles within the ciliary body contract to flatten the lens, by putting tension upon the suspensory ligament, when the eye is focusing upon distant objects. Conversely, circular muscles can contract allowing the ligament to slacken, causing the elastic lens to adopt a rounder shape, as when the eye is focusing on near objects. This is known as accommodation.

Lepidoptera an order of ENDOPTERYGOTE insects comprising the moths and butterflies. The body, wings and legs of adults are covered with numerous minute scales and the mouth parts are modified to form a long tube, the PROBOSCIS, which is often held coiled beneath the head. This is uncoiled and extended to enable the insect to feed on nectar. Two large pairs of membranous wings, which may be brightly coloured, are present. In the butterflies the resting wings are held vertically. These insects are active by day and tend to have filamentous, club-tipped antennae. In the moths, nocturnal forms occur and the resting wings may be held in various positions and the antennae may be feathery. The larvae of lepidopterans are known as caterpillars. They have a well-defined head with chewing mouthparts and a segmented body, usually with each segment bearing a pair of legs. Many caterpillars are serious economic pests of food crops and trees and may transmit plant diseases. The caterpillar becomes an adult by undergoing METAMORPHOSIS via a PUPA (or chrysalis). The pupa may be surrounded by a silk cocoon, which the caterpillar produces from modified salivary glands, the silk glands. Others use leaves or similar material to construct a cocoon. Examples of the order are clothes moths, flour moths, gypsy moths and silkworm moths, swallowtail and monarch butterflies.

leprosy an infectious disease that affects the skin, nerves and mucous membranes of the patient. The symptoms of leprosy include severe lesions of the skin and destruction of nerves, which can lead to disfigurements such as wrist-drop and claw-foot. Leprosy is caused by the airborne bacterium, *Mycobacterium lepra*, and, fortunately, is not highly contagious as transmission involves direct contact with this bacterium. The likeliest source of infection, therefore, arises from the nasal secretions (swarming with bacteria) of patients and not from the popular misconception of touching the skin of an infected individual. Leprosy is curable, and the treatment, using sulphone drugs, has the beneficial side-effect of making the patient non-infectious, even if he or she is not completely cured. Although the incidence of leprosy was once worldwide, it is now mostly confined to tropical and subtropical regions.

lethal gene a gene that, if expressed, will cause the death of the individual. The fatal effect of the expressed gene usually occurs in the prenatal developmental stage of the individual, i.e. the embryonic stage for animals and the pupal stage for insects. Although most examples of lethal mutants fail to survive to adulthood, there is one well-researched genetic disorder, called Huntington's chorea, that does not usually affect the individual until middle age. Huntington's chorea is caused by a single dominant gene, and thus half the children of an affected parent will inherit the genetic disorder, although fortunately this disease is rare.

leucocyte a large, colourless cell formed in the bone marrow and subsequently found in the blood of all normal vertebrates. It is commonly known as a white blood cell and plays an important role in the IMMUNE SYSTEM of an individual. Leucocytes are produced in the bone marrow, spleen, thymus and lymph nodes of the body and can be classified into the following three groups in order of decreasing constituency of leucocytes:

Group	%	Functions
Granulocyte	70	Helps combat bacterial and viral infection and may also be involved in allergies.
LYMPHOCYTE	25	Destroys any foreign bodies either directly (T-CELLS) or indirectly by producing antibodies (B-CELLS).
Monocyte	5	Ingests bacteria and foreign bodies by the mechanism called PHAGOCYTOSIS.

leucoplast a colourless object that contains starch and is found in some plant cells. If a leucoplast contains the pigment CHLOROPHYLL it may develop into a CHLOROPLAST.

leukaemia a cancerous disease in which there is an uncontrolled proliferation of white blood cells (LEUCOCYTES) in the bone marrow. The white blood cells fail to mature to adult cells and thus they cannot function as an important part of the defence system against infections. Although the definite cause of leukaemia is as yet unknown, there is growing suspicion that certain viruses may cause it and that perhaps there is a hereditary component. Unfortunately, leukaemia is not a curable disease, but there are methods effective in suppressing the reproduction of white blood cells—radiotherapy and, more commonly, chemotherapy. These methods bring the disease under control and thus help prolong the patient's life.

LH *abbreviation for* LUTEINIZING HORMONE.

lichen a plant-like growth that is formed from two

organisms existing in a symbiotic relationship (*see* SYMBIOSIS), namely a FUNGUS and an ALGA. The fungus usually belongs to the subdivision Ascomycotina or Basidiomycotina and the alga is either a blue-green alga (CYANOBACTERIA) or a green alga (CHLOROPHYTA). A lichen forms a distinct structure, which is not similar to either partner alone. Usually most of the plant body is made up of the fungus with the algal cells distrbuted within, either evenly dispersed throughout the HYPHAE or more commonly as a thin layer. The fungus protects the algal cells, and the alga provides the fungus with food through PHOTOSYNTHESIS. A lichen is typically very slow growing and varies in size from a few millimetres to several metres across. It may occur as a thin, flat crust or bear leaf-life lobes, or be upright and branching. Frequently they occur in conditions that are too cold or exposed for other plants, such as arctic and mountainous regions. Some are epiphytes (use another plant for support) especially growing on tree trunks. Non-sexual reproduction occurs with the production of structures containing components from both partners (soredia) or by fragmentation. Sexual reproduction is confined to the fungus producing spores, which, when they germinate, can only survive if the algal partner is present. Most species are also sensitive to air pollution (especially sulphur dioxide) and have been used as indicators. Reindeer moss and Iceland moss are lichens that provide an important food source in arctic regions. Other lichens provide substances that are used in dyes, perfumes, cosmetics and medicines as well as poisons.

life cycle the total sequence of events undergone by

individuals belonging to a particular species from fusion of GAMETES in one generation to the same stage in their offspring. In many animals, and in most of the higher forms, the germ cells contained within the parents' reproductive organs produces gametes by undergoing MEIOSIS. The two gametes fuse at fertilization and the ZYGOTE so formed develops into an individual that resembles its parents. However, variations in this pattern can occur, especially in simpler organisms (*see* ASEXUAL REPRODUCTION and PARTHENOGENESIS). In plants, however, meiosis results in the formation of spores that develop into plants that are known as the GAMETOPHYTE generation, i.e. they bear the sex organs that produce the gametes. This may be very different from the stage in the life cycle that produces the spores, the SPOROPHYTE generation, and one or other stage may be reduced or dominant depending upon the species. The sporophyte generation is established again when the gametes produced by the gametophyte generation fuse. The sporophyte is the dominant form in the life cycle of seed plants with the gametophyte being very much reduced. The gametophyte is dominant in bryophytes, and the sporophyte is dependent upon it (for nutrition, etc) to a greater or lesser extent. The gametophyte is HAPLOID, and the sporophyte is DIPLOID but produces haploid spores.

ligament a kind of CONNECTIVE TISSUE, formed chiefly of COLLAGEN, which is found between the movable JOINTS of vertebrates. Ligaments join one bone to another and restrict movement, hence preventing dislocation.

ligand any MOLECULE or ATOM capable of forming a

bond with another molecule (usually a metallic CATION) by donating an ELECTRON PAIR to form a complex ION. In biological terms, ligand refers to any molecule capable of binding with a specific ANTIBODY.

ligase any of a number of enzymes that catalyse the condensation of two molecules using energy released from the splitting of ATP. Ligases are very important enzymes biologically in that they are involved in the repair of DNA and other molecules, and also in synthesis. They are widely used in GENETIC ENGINEERING to insert a foreign DNA sequence into another one.

light reactions the biochemical processes that generate ATP, oxygen and a reduced coenzyme called NADPH during PHOTOSYNTHESIS in the presence of light. The light-dependent reactions occur in the inner membranes of CHLOROPLASTS and require water and several forms of the pigment CHLOROPHYLL. Two of the products of the light reactions, ATP and NADPH, enter the CALVIN CYCLE, whereas the third product, OXYGEN, is released by the plant.

lignin a complex organic polymer, which is associated with CELLULOSE, that is laid down within the cell walls of plants during secondary thickening. It makes the cell walls woody and rigid and forms 20–30 per cent of wood, being the most common plant polymer after cellulose.

lignite *or* **brown coal** a coal that in rank falls between peat and bituminous coal. It shows little alteration, the woody structure often being visible. It has a high moisture content that can be reduced, and it is found in Tertiary basins in Britain, and in Europe.

limestone a SEDIMENTARY ROCK composed mainly of calcite with dolomite, which may be organic, chemical or detrital in origin. There is a tremendous variety in the make-up of limestones, which may comprise remains of marine organisms (corals, shells, etc); minute organic remains (as in CHALK); grains formed as layered pellets (*called* ooliths) in shallow marine waters and calcareous muds. Recent deposits of calcium carbonate are found in shallow tropical seas. Modification after deposition is usually extensive, and both the composition and structure may change due to compaction, recrystallization and replacement. Limestone has many uses commercially including building stone, roadstone and aggregate, as a source material in the chemicals industry, and in their natural state as aquifers and petroleum reservoir rocks.

limiting reactant any substance that limits the quantity of the product obtained during a chemical reaction. The limiting reactant can be identified using the chemical equation of a reaction as it will be the smallest quantity in comparison to the other reactants and products.

linkage two or more GENES are said to be linked if they occur together on the same CHROMOSOME, and hence the characteristics that they provide for tend to be inherited together. This means that linked genes will pass together into nuclei formed as a result of MEIOSIS. Linkage can be broken by a process known as crossing over, which occurs at a stage during meiosis. This involves a breaking and rejoining of chromosome strands so that new combinations of genes are produced and the alteration of characteristics becomes possible.

Linnaeus C—Carolus Linnaeus, a Swedish physician and naturalist (1707–88) who devised the modern system of classification known as binomial nomenclature. This ascribes an organism into a genus and species so that each has a double biological name. In addition, he arranged organisms into a hierarchical series of categories (*see* CLASSIFICATION), and this is known as the Linnaean system (*see also* SPECIES).

linoleic acid a polyunsaturated, liquid, essential fatty acid widely found in many plant fats and oils, e.g. soya bean oil and ground nut oil. It has two double bonds in its structure and comprises about 20 per cent of the total fatty acid content in the triglycerides of mammals (*see* ESSENTIAL FATTY ACID).

linolenic acid a polyunsaturated, liquid, essential fatty acid with three double bonds in its structure. It is found in algae and also in some plant oils, e.g. soya bean oil and linseed (*see* ESSENTIAL FATTY ACID).

lipase any enzyme capable of breaking down fat to form FATTY ACIDS and GLYCEROL. Lipases function in alkaline conditions and are most abundant in the pancreatic secretions during digestion.

lipids an all-embracing term for oils, fats, waxes and related products in living tissues. They are ESTERS of FATTY ACIDS and form three groups: simple lipids, including fats, oils and waxes; compound lipids, which includes PHOSPHOLIPIDS; and derived lipids, which includes STEROIDS.

lipolysis the hydrolysis (breakdown) of stored lipids, particularly triglycerides in fats and oils, within an organism. The stored fat reserves provide a source of energy, which may be needed, for example, during starvation. LIPASE enzymes break down the triglycerides into GLYCEROL and FATTY ACIDS, and these are then transported to the tissues where they are utilized.

lipoprotein (LDL) any protein that has a FATTY ACID as a side chain. Lipoproteins have significant importance in certain biological processes as they function as a transport mechanism for essential molecules. For example, as CHOLESTEROL is extremely hydrophobic, there would be no method of transporting it to its target body tissues. This problem is solved by low-density lipoprotein surrounding the cholesterol molecule and forming a hydrophilic molecule that can be transported by body fluids.

liposome a spherical vesicle surrounded by a membrane, which is artificially created in the laboratory and in structure and function is like a cell organelle. Liposomes are constructed by adding an aqueous solution to a phospholipid gel. The properties of liposome membranes are studied to gain insight into the function of living cell membranes. Liposomes are also used medically to transport quite toxic drugs as they can be introduced into living cells. The liposomes safely contain the drug (in the blood) until they reach the diseased tissue which is being targeted where the membranes can be induced to melt. This is applicable in certain forms of cancer treatment.

lipotrophin produced in the anterior PITUITARY GLAND, one of two peptide HORMONEs that are involved in controlling the breakdown of fat deposits and the transfer of lipids into the bloodstream.

liquefaction of gases a gas may be turned into its liquid form by cooling below its critical temperature (the temperature above which a gas cannot be liquefied by pressure alone). In addition, pressure may be required. For gases such as oxygen, helium and nitrogen, low temperatures are used.

liquid a fluid state of matter that has no definite shape and will acquire the shape of its containing vessel as it has little resistance to external forces. A liquid can be regarded as having more KINETIC ENERGY than a solid but less kinetic energy than a GAS. It is considered that the average kinetic energy will increase as the temperature of the liquid rises.

lithosphere that layer of the Earth, above the asthenosphere, that includes the crust and the top part of the mantle down to 80–120 kilometres in the oceans and around 150 km in the continents. The base is gradational and varies in position depending upon the tectonic and volcanic activity of the region. The lithosphere comprises the blocks that constitute plate tectonics.

litmus (*see* INDICATOR) a natural compound obtained from lichens which is used as an indicator, turning red to indicate acid conditions and blue for alkaline.

litre a unit of volume given the symbol l and equal to 1000 cubic centimetres, i.e. $1l = 1000cm^3 = 1dm^3$. One gallon is approximately 4.5 litres.

littoral a term for the shallow water environment of a lake or the sea that lies close to the shore. In the sea, the littoral zone includes the tidal area between the high and low water marks. It is characterized by the penetration of light so that rooted aquatic plants can grow and there is usually an abundance and diversity of organisms.

liver a very important large organ, present in the abdomen of vertebrates, that plays a critical regulatory role in many metabolic processes. The hepatic portal vein conveys the products of digestion to the liver where excess GLUCOSE is converted to GLYCOGEN, which is then stored as a food reserve. Excess amounts of amino acids are converted to ammonia and then to UREA, which is excreted by the kidneys. LIPOLYSIS takes place in the liver as does the production of bile which is stored in the GALL BLADDER. Some poisons are broken down (detoxified) in the liver, e.g. alcohol, and various plasma proteins and blood-clotting substances such as FIBRINOGEN, prothrombin and ALBUMIN are synthesized. Iron is stored in the liver and STEROID hormones are converted to CHOLES-TEROL. VITAMIN A is synthesized and stored, the liver also being a storage site for VITAMINS D, E and K. The removal of (damaged) red blood cells also occurs in the liver.

loam a type of soil with sand, silt and clay in roughly equal proportions, often with some organic matter.

locus the name given to the region of a CHROMOSOME occupied by a particular GENE.

loess a sediment formed from the aeolian (i.e. with wind as the agent) transportation and deposition of mainly silt-sized particles of quartz. It is well sorted, unstratified, and highly porous, and although it can maintain steep to vertical slopes, is readily re-worked. Loess is widespread geographically, and although thicknesses of Chinese deposits exceeds 300 metres, it is normally only a few metres. Its origin has been hotly debated, but it is now accepted that the loess particles are produced by glacial grinding, frost cracking, hydration in desert regions and aeolian impact of sand grains. The wind is the primary and essential agent in the process.

lone pair a pair of electrons that are not shared by

another atom but that can form co-ordination compounds.

longshore drift the movement of material (sand and shingle) along the shore by a current parallel to the shore line (i.e. a longshore current). Longshore drift occurs in two zones: beach drift, at the upper limit of wave activity (i.e. where a wave breaks on the beach) and the breaker zone (where waves collapse in shallow water) where material in suspension is carried by currents.

lophophore an organ used for suspension feeding that is found in three invertebrate phyla: BRYOZOA, PHORONIDA and BRACHIOPODA. It comprises a circular fold of the body wall bearing ciliated tentacles and surrounds the mouth. The CILIA beat and draw water in towards the mouth, helping to trap food particles.

Lotka's equations mathematical expressions used in ecology for the relationships between two species competing for food or space in the same area, and the situation when one is predator, and one prey.

LTH *abbreviation for* luteotropic hormne (*see* PROLACTIN).

lumbar vertebrae vertebrae that occur above the sacral vertebrae and below the thoracic vertebrae in the lower back. They bear processes for the attachment of back muscles in the mammals.

lumen the space or cavity enclosed by any sac-like organ, duct or tube such as those of the ALIMENTARY CANAL and blood vessels.

luminescence the emission of light by a living organism that is not a consequence of raising the body temperature of the organism. For luminescence to occur, the cells of the organism must contain the protein luciferin, the enzyme luciferase, and the energy source for the reaction, ATP.

lung the sac-like organ of all air-breathing vertebrates, which serves for respiration. In mammals a pair of lungs is situated within the rib cage in the thorax. Each lung consists of a thin, moist, much folded membrane, the folding serving to increase the surface area through which gaseous exchange can occur. Air enters the body through the windpipe or TRACHEA, which branches into two BRONCHI , each serving one lung. Further branching of the bronchus occurs, eventually forming numerous tiny bronchioles. Each bronchiole is surrounded by a tiny sac, an ALVEOLUS, formed from a fold of the lung membrane that is the actual site of gaseous exchange. The exchange of oxygen and carbon dioxide occurs between the numerous blood CAPILLARIES that serve the alveoli on one side of the membrane and the air on the other side. Respiratory movements of the DIAPHRAGM in mammals ventilate the lungs, which do not have muscles of their own. In birds the lungs are small and air sacs are the main site for gaseous exchange. The lungs are served by the PULMONARY ARTERIES and PULMONARY VEINS.

luteinizing hormone (LH) a gonadotrophin hormone that is produced by the anterior PITUITARY GLAND and acts upon the male and female sex organs. The anterior pituitary produces luteinizing hormone under the influence of releasing factor secreted by the HYPOTHALAMUS. Luteinizing hormone stimulates ovulation in the female ovary and the formation of the CORPUS LUTEUM and PROGESTERONE production. In the male TESTES secretory interstitial cells are stimulated to produce testosterone (*see*

ANDROGEN). This hormone is also known as *interstitial-cell-stimulating hormone* or *ICSH*.

luteotropic hormone *see* **prolactin**.

lymph a colourless, watery fluid that surrounds the body cells of vertebrates. It circulates in the LYMPHATIC SYSTEM but is moved by the action of muscles, as opposed to the contraction of the heart. Lymph consists of 95 per cent water but contains protein, sugar, salts, and LEUCOCYTES, and transports fat from the gut wall during digestion.

lymphatic system a network of tubules found in all parts of a vertebrate's body except, if present, the central nervous system. The tubules drain the body fluid, LYMPH, from the tissue spaces, and as they gradually unite to form larger vessels, the lymph vessels finally drain into two major lymphatic vessels that empty into veins at the base of the neck. The system consists of vessels and nodes, where the fluid is filtered, bacteria and other foreign bodies are destroyed, and LYMPHOCYTES enter the lymphatic system. The lymphatic system is not only an essential part of the IMMUNE SYSTEM, but also carries excess protein and water to the blood and transports digested fats.

lymphocyte a cell that is produced in the bone marrow and is an essential component of the IMMUNE SYSTEM as it will differentiate into either a B-CELL or a T-CELL. Lymphocytes are a form of white blood cell (LEUCOCYTE), which collect in the lymph nodes of the lymphatic system and defend the body against foreign bodies and bacterial infections. Lymphocytes are also present in blood but in a smaller percentage.

lymphoid tissue a type of tissue found in vertebrates, responsible for the production of LYMPHOCYTES and antibodies. Examples are the embryonic liver, thymus gland, bone marrow, spleen and lymph nodes.

lysis the destruction of cells, commonly BLOOD CELLS, by antibodies called **lysins**. More generally, the destruction of cells or tissues by pathological processes, e.g. autolysis, where cells are broken down by enzymes produced in the cells undergoing breakdown.

lysosome a sac-like organelle surrounded by a membrane, which is contained within the CYTOPLASM of animal cells. Various types of lysosome occur, and they contain different sorts of digestive enzymes. These are responsible for the digestion of substances contained in food vacuoles, the breakdown of foreign material such as bacteria that may enter the cell and, when a cell dies, the dissolution of all cellular material.

lysozyme an enzyme that is present in tears, nasal secretions and on the skin, and that has an antibacterial action. It breaks the cell wall of the bacteria leaving them open to destruction. Lysozyme also occurs in egg-white, and it was the first enzyme for which the three-dimensional structure was determined.

M

macrophage a specialized cell that forms part of the IMMUNE SYSTEM of vertebrates. Macrophages are derived from MONOCYTES in the blood system and can move to infected or inflamed areas of the body using PSEUDOPODIA. In such areas, they ingest and degrade broken cells and other debris including microbes, by means of PHAGOCYTOSIS.

macula (*plural* **maculae**) sensory cells in the INNER EAR that are instrumental in the maintenance of balance and provide information about the head's position whether the animal is moving or at rest. Also, an area around the fovea (the point in the RETINA with no ROD CELLS but many CONE cells), which is the centre of the visual field.

major histocompatibility complex (MHC) a gene complex comprising a collection of GLYCOPROTEIN molecules in the plasma membrane of cells that are important in the IMMUNE SYSTEM, in the recognition of self and non-self (i.e. distinguishing between the body's own cells and foreign molecules or ANTIGENS). There are two main groups—class I and II MHC molecules, the former marking nucleated cells (most of the body's cells). The latter, class II, mark MACROPHAGES and B LYMPHOCYTES and fulfil an important role between cells of the immune system, including the HISTOCOMPATIBILITY ANTIGENS. The MHC in humans is on chromosome 6 and includes the HLA SYSTEM, and is unique to each individual.

Malacostraca the largest group (subclass) of the class CRUSTACEA that includes the DECOPODA (crabs and lobsters), prawns and shrimps (Amphipoda) and woodlice (Isopoda).

malaria an infectious disease caused by certain parasites in the blood of the victim. Malaria is transmitted by an infected mosquito biting a human, thus injecting the parasite from the salivary gland of the mosquito into the bloodstream of the human. After the parasites have developed in the victim's liver, they are released into the bloodstream and attack red blood cells. Early symptoms of malaria include headaches, body aches and chills. As the disease progresses, malarial attacks are frequent and cause sickness, dizziness and sometimes delirium in the victim; the attacks seem to coincide with the bursting of infected red blood cells. Fortunately, there are many highly effective drugs for treating malaria.

malic acid ($C_4H_6O_5$) a crystalline acid that in its l-form forms an intermediate metabolite in the KREBS CYCLE and in certain plants in PHOTOSYNTHESIS. It occurs in many acid fruits, such as apples, grapes and gooseberries.

malleus *or* **hammer** the first of the EAR OSSICLES in the MIDDLE EAR of mammals.

Malpighian corpuscle (or body) the part of the NEPHRON in the vertebrate KIDNEY that comprises the GLOMERULUS and its cup-shaped end.

maltose malt sugar, a disaccharide formed from two

linked GLUCOSE molecules formed by the breakdown of starch by the enzyme amylase (*see* DIASTASE). Maltose occurs in germinating seeds, and the malt used in the brewing industry is from germinating barley seeds that have then dried.

Mammalia a vertebrate class that contains approximately 4500 species. Mammals have a number of distinctive characteristics including hair, HETERODONT teeth and sound conducting EAR OSSICLES in the MIDDLE EAR. They are endothermic, i.e. warm-blooded (*see* HOMOIOTHERMY), and sweat gland secretions cool the skin. The four-chambered HEART keeps oxygenated and deoxygenated blood separate, and the muscular DIAPHRAGM assists in VENTILATION.

MAMMARY GLANDS secrete milk for the young and fertilization is internal, the egg developing into an embryo within the female. The placenta of certain mammals lines the UTERUS and provides a membrane through which nutrients diffuse into the blood of the embryo, and waste diffuses out. Mammals have highly developed sense organs and larger brains than other vertebrates of a comparable size. The capacity to learn and rationalize has enabled mammals to colonize and live in almost all habitats.

There are three subclasses: EUTHERIA, the placental animals; METATHERIA, the marsupials and PROTOTHERIA, the monotremes. The monotremes are the most primitive, being the only mammals that lay eggs.

Mammals evolved from the reptiles roughly 220 million years ago in the Triassic, and the therapsids (mammal-like reptiles) formed the "ancestral stock." Although the therapsids disappeared during the time of the dinosaurs, offshoots following the mammalian line co-existed with the dinosaurs through the Mesozoic (*see* APPENDIX 5).

mammary glands glands, peculiar to female animals, that secrete milk to nourish young and probably owe their origin to modification of sweat glands. The number varies according to species, and rudimentary versions may be present in the male. The gland usually opens as a nipple or teat, each supplied with a number of, or single (respectively), milk ducts.

mandible the lower jaw of vertebrates. Also, a horny mouthpart, occurring in pairs and used by insects, crustaceans, centipedes and millipedes to bite and crush the food.

map unit the region between two gene pairs on the same CHROMOSOME, where one per cent of the possible products during MEIOSIS are recombinant.

mass the measure of the quantity of matter that a substance possesses. Mass is measured in grams (g) or kilograms (kg).

mass number (A) the total number of PROTONS and NEUTRONS in the nucleus of any atom. The mass number therefore approximates to the RELATIVE ATOMIC MASS of an atom.

mast cell a large, blood-borne cell that has a fast-acting role in the IMMUNE SYSTEM. In allergies, mast cells will be triggered to release HISTAMINE when the IMMUNOGLOBIN, IgE, has already attached itself to the foreign body that has entered the body.

mastoid process a small extension from the temporal bone (in the skull) that contains air spaces and communicates with the MIDDLE EAR.

matter any substance that occupies space and has mass: the material of which the universe is made.

maxilla (*plural* **maxillae**) in vertebrates, one of the upper jaw bones that bear teeth which in mammals is all the teeth except the incisors. Also, in insects, crustaceans, centipedes and millipedes, one of a pair of mouthparts that are behind the MANDIBLES. The maxillae assist in feeding, and in crustacea the second pair fuse to form the labium (a lower lip).

meatus a general term for a small passage or canal in the body.

mediastinum the space between the lungs that contains the heart, trachea, oesophagus etc.

medulla the central part or tissues of an organ, where the outer portion is usually called the CORTEX. In plants, the PARENCHYMA tissue in the centre of items, up to the vascular tissue.

medulla oblongata a part of the vertebrate BRAINSTEM that regulates the control of blood pressure, heart beat and similar involuntary actions. It connects to many of the CRANIAL NERVES and is derived from the HINDBRAIN.

medullary ray plates of PARENCHYMA cells in the vascular tissue of plant roots and stems. The rays act as food stores and transport nutrients. The plates vary in thickness from one to several cells and may be primary or secondary depending upon their location or formation.

medullary sheath *see* **myelin sheath**.

medusa (*plural* **medusae**) a stage in the life cycle of coelenterates (*see* CNIDARIA) that is free-swimming. Some cnidarians exist only as POLYPs and some as medusae, while others pass through both forms in their life cycle. The medusa is essentially a flattened form of polyp that is umbrella-shaped, with a mouth in the centre of the lower surface. Tentacles that ring the mouth (also used as the anus) are used to catch prey and push it into the mouth. Medusae move in the water by drifting and by contractions of the body. *Hydra* alternates between polyp and medusa stages, while the jellyfish has the medusa dominant.

meiosis a chromosomal division that produces the germ cells (GAMETES in animals and some plants, sexual spores in fungi). Meiosis involves the same stages as MITOSIS, but each stage occurs twice, and, as a consequence, four HAPLOID cells are produced from one DIPLOID cell. Meiosis is an extremely important aspect of sexual reproduction, as the production of haploid cells ensures that during FERTILIZATION the chromosomal number is constant for every generation. It also gives rise to genetic variation in the daughter haploid cells by the rearrangement and GENETIC RECOMBINATION if genetic variability already exists in the parent diploid cell.

melanin a dark pigment responsible for colouring the skin and hair of many animals, including humans. Differences in skin colour are due to variations in the distribution of melanin in the skin and not to differences in the number of cells, **melanocytes**, that produce the pigment.

melanism when some individuals in a population are black due to overproduction of the pigment MELANIN. It is particuarly well noted with insects found in some industrial areas where, due to the increase in industrial pollution, a species evolved to compensate. In this case of *industrial melanism*, the peppered moth was found to be much darker in polluted areas compared to the non-polluted parts, thus making it less visible to a predator.

melting point the temperature at which a substance is in a state of equilibrium between the solid and liquid states, e.g. ice/water. At 1 atmosphere pressure, the melting point of a pure substance is a constant and is the same as the freezing point of that substance. At constant pressure, the melting point of a substance is lowered if it contains impurities (the reason for adding salt to make ice melt). Although an increase in pressure lowers the melting point of ice, in most substances increased pressure will raise their melting points.

membrane a tissue, occurring in a sheet-like form, that covers, lines or joins cells, organelles, organs etc. Membranes consist of a double layer of LIPIDS (a *bilayer*), mainly the PHOSPHOLIPIDS lecithin and cephalin, with PROTEIN molecules. Movement of ions across a membrane is achieved by carriers (as in ACTIVE TRANSPORT), and while water and fat-soluble substances can pass, sugars cannot.

membrane potential the difference in electric potential between the inside and the outside of the plasma membrane of all animal cells. Membrane potentials exist due to the different ionic concentrations within and outwith the cell, and also the selective permeability of the plasma membrane to specific ions (notably potassium [K^+], sodium [Na^+] and chloride [Cl^-] ions). When measured using a microelectrode, the resting potential of most muscle and nerve cells is -60mV on the inside of the plasma membrane.

Mendel's laws of genetics laws of heredity deduced by Gregor Mendel (1822–84), an Austrian monk, who discovered 16 basic rules in experiments with generations of pea plants. He established that a trait, such as flower colour or plant height, had two factors (hereditary units) and that these do not blend but can be either dominant or recessive. Without knowledge of genes or cell division, he developed the following laws of particular inheritance:

First law—each factor segregates from each other into the GAMETES. It is now recognized that ALLELES are present on HOMOLOGOUS CHROMOSOMES, which separate during MEIOSIS.

Second law—factors for different traits undergo INDEPENDENT ASSORTMENT.

During gamete formation, the segregation of one gene pair is independent of any other gene pair unless the genes are linked (*see* LINKAGE).

meninges (*singular* **meninx**) three membranes that cover the spinal cord and brain. The *pia mater* is the innermost and supplies capillaries to the ventricles of the brain, then the *arachnoid*, and the outer membrane is the *dura mater*. Between the first two are arachnoid spaces, filled with fluid, where the cerebrospinal fluid is reabsorbed into the venous system. The dura mater is thick and lines the skull, and the dural sinus (between the dura mater and the arachnoid) drains blood from the brain.

menstrual cycle the cycle in female humans that replaces the OESTROUS CYCLE of other mammals. It is approximately monthly and associated with OVULATION. Ovulation occurs when the endometrium (uterus lining) thickens with more blood vessels, in preparation for the possible implantation of an embryo. Ovulation occurs at the middle of the cycle and if fertilization does not occur the lining of the uterus breaks down and is discharged from the body during *menstruation*.

meristem a region of plant tissue where there is active cell division. The cells thus formed are modified to become different tissues whether epidermis or cortex etc. Particularly important are stem and root tip meristems (apical).

mesencephalon *see* **midbrain**.

mesenchyme embryonic MESODERM that gives rise to blood, cartilage, bone etc.

mesentery a thin tissue in sheet form, occurring as a double-layered extension of the PERITONEUM, that attaches to and supports the organs in an animal's body cavity. In vertebrates the mesentery contains blood vessels and nerves to the gut. Mesenteries also support the reproductive organs.

mesoderm the middle GERM LAYER of the GASTRULA that lies between the ENDODERM and ECTODERM. From it (in mammals) develop the musculature, cartilage and bone, blood and other parts of the circulatory system, the excretory system and reproductive ducts.

mesoglea *see* **Cnidaria**.

mesophyll the internal tissues of a leaf that are between the upper and lower epidermal layers. Mesophyll tissue contains CHLOROPLASTS, which are concentrated at a site that allows maximum absorption of light for PHOTOSYNTHESIS.

mesophyte a term referring to plants that grow in soil well supplied with nutrients and water and, as such, are not able to conserve water. Thus in very dry or drought conditions, they will wilt very easily.

mesothelium a layer of thin, platelike cells, derived from the mesoderm, that is similar to epithelium. The cells cover the inside of the abdominal cavity and surround the heart to form part of the PERITONEUM and PLEURA.

messenger RNA (mRNA) a single-stranded RNA molecule that contains ribose sugar, phosphate and the following four bases—ADENINE, CYTOSINE, GUANINE, and URACIL. Messenger RNA has the important role of copying genetic information from DNA in the nucleus and carrying this information in the form of a sequence of bases to RIBOSOMES in the cell, where TRANSLATION occurs to synthesize the specified protein that was encoded in the nuclear DNA.

metabolism the chemical and physical processes occurring in living organisms, which comprise two parts: CATABOLISM and ANABOLISM. ENZYMES control metabolic reactions, which tend to be similar throughout the plant and animal kingdoms.

metabolite a substance that takes part in the process of METABOLISM whether originating from outside the organism or produced within, in a biochemical reaction.

metacarpus in terrestrial vertebrates, the hand (the palm of man), that consists of a number of bones (metacarpals) articulating with the wrist (CARPUS) and the fingers (PHALANGES). The PENTADACTYL LIMB has five metacarpals but there is variation with some species having fewer.

metamerism or **metameric segmentation** or **segmentation** the division of an animal's body into segments, as in the ANNELIDA. Each segment has the same organs repeated, i.e. muscle, blood vessels, nerves, and in the annelids and arthropods the division is obvious both externally and internally. Metamerism also occurs in the embryonic development of all vertebrates although only in certain organs or tissues, e.g. muscle, and there is no external sign.

metamorphosis the rapid change that an animal undergoes from the larval to the adult form. It occurs in the development of many invertebrates and some amphibians, e.g. butterflies and frogs respectively. The changes are controlled by hormones and in many cases there is significant breakdown of larval tissues and the adult grows through the division and differentiation of cells that were inactive in the larva.

In the *incomplete* metamorphosis of, for example, grasshoppers, the young are very similar to the adults save for being smaller and the transition is achieved through a series of moults. Insects that undergo *complete* metamorphosis have larval stages that are different from the adult.

metaphase a stage in MITOSIS or MEIOSIS in cells of EUCARYOTES. During metaphase, the chromosomes are organized and attached to the equator of the spindle by their CENTROMERES. Metaphase occurs only once in mitosis but twice in meiosis.

metastasis the process of malignant cancerous cells spreading from the affected tissue to create secondary areas of growth in other tissues of the body.

metatarsus in terrestrial vertebrates, the foot, the hindlimb equivalent of the METACARPUS. In this case, bones (metatarsals) articulate with the ankle (tarsus), and the toe bones (PHALANGES).

Metatheria one of the three mammalian sub-classes (*see* MAMMALIA), the marsupials. The distinctive feature is that the female has an abdominal pouch, or marsupium, in which the young complete their development. Marsupials are born at a very early stage of their development and after birth crawl into the pouch and fix on a teat.

Marsupials include the kangaroos, koala bears and opossums, and are restricted to Australia and America. The group evolved during the late CRETACEOUS, about 80 million years ago and after the break-up PANGAEA, Australia and South American became island continents, which restricted the spread of marsupials. As a result, and particularly in Australia, the marsupials have adapted to fill niches that might otherwise have been occupied by placental mammals.

Metazoa a sub kingdom characterized by multicellular animals with grouping of cells into tissues, with co-ordination by a nervous system. Included are all the vertebrates, and invertebrates save for the PROTOZOA and sponges (PORIFERA).

methane the first member of the HOMOLOGOUS SERIES of ALKANES. It is a colourless, odourless gas with the chemical formula CH_4. Methane is the main constituent of coal gas and is a byproduct of any decaying vegetable matter. It is flammable and is used in industry as a source of hydrogen.

methyl (or Me) the group CH_3.

methylation the process of adding a METHYL group to a compound. In biology it is the addition of a methyl group to a NUCLEIC ACID base, e.g. ADENINE or CYTOSINE.

methylene blue a blue dye that is used in microscopy to stain the nuclei of animal tissues.

MHC *abbreviation for* MAJOR HISTOCOMPATIBILITY COMPLEX.

micelle a term used to describe a submicroscopic aggregate of molecules in a colloidal solution (*see* COLLOID). It is applicable especially to soaps and dyes.

microbe *see* **micro-organism**.

microbiology the study of micro-organisms, bacteria, fungi, viruses. Their study has developed, from aspects of disease to, in particular, their genetics. This is because microbes can be readily and rapidly cultured in the laboratory and used to study genetic and biochemical processes.

micro-organism *or* **microbe** an organism that can be seen only with a microscope. Included are bacteria, viruses, protozoans, some algae and fungi.

microphagy the general term for feeding methods used by animals that feed on particles that are very small in relation to their own size. Because of the size of the food particles, feeding tends to occur continually. A number of devices are adopted to enable sieving of the particles from water, e.g. baleen plates (*see* WHALEBONE) in baleen whales, also gills in bivalves and lophophores in brachiopods.

microscopy the study of samples using one of a number of microscopes e.g. light microscope, ELECTRON MICROSCOPE. For light microscopes tissue samples must be prepared in a specific way (*see* FIXATION and STAINING) and very thin slices taken for study (*see* MICROTOME).

Samples for study with the electron microscope are embedded in Araldite after fixing and prior to sectioning.

microtome a machine used to slice very thin sections of plant or animal tissue for study with the microscope. After fixing, samples are frozen or embedded in paraffin wax and sections then cut. Samples for light microscopy are cut to 3-10μm but electron microscopy requires thicknesses of 20–100 nm.

microtubule a long, hollow fibre of PROTEIN that is

found in all higher plant and animal cells. Microtubules have different functions in different cells, e.g. they form the spindle during MITOSIS, give strength and rigidity to the tentacles of some unicellular organisms, and they are also found in parts of nerve cells.

microvillus (*plural* **microvilli**) tiny, finger-like extensions from the free surfaces of epithelial cells, especially surfaces involved in absorptive or secretory functions, e.g. in the small intestine. Thousands of microvilli form a *brush border* and provide a much increased surface area for absorption. Each microvillus is about 1 μm long.

midbrain *or* **mesencephalon** in the vertebrate embryo, one of three parts of the brain. However, unlike the FOREBRAIN and HIND BRAIN, the midbrain does not develop additional zones. In mammals, it becomes part of the BRAINSTEM.

middle ear in vertebrates, the cavity between the inner and outer ear that contains the EAR OSSICLES. It is an air-filled cavity linked via the EUSTACHIAN TUBE to the pharynx.

mid gut the central section of the alimentary canal of vertebrates that includes most of the SMALL INTESTINE and is involved in digestion of food and absorption of nutrients.

migration the seasonal movement of animals, especially birds, fish, and some mammals (e.g. porpoises). Climatic conditions tend to be the trigger where perhaps lower temperatures mean less available food. Some animals, and particularly birds, travel vast distances, e.g. the golden plovers fly 8000 miles from the Arctic to South America. Migrating animals seem to use three mechanisms for finding their way. Over short distances, an animal moves to successive, familiar landmarks (*piloting*). In *orientation*, a straight line path is taken, based upon the animal adopting a specific (compass) direction. *Navigation* is the most complex process because it necessitates taking a direction relative to, and having determined, the animal's present position. It now seems that some species of birds use the sun, stars (often the North star which moves very little) and an internal clock to compensate for the relative positions of these bodies. If cloud cover should obscure the sun, many species are able to continue their migration quite accurately, probably because they can orient themselves with respect to the Earth's magnetic field.

milk an aqueous fluid secreted by MAMMARY GLANDS for suckling young. In addition to water, which is the main constituent, milk contains lipids, protein (mainly CASEIN) and sugar (LACTOSE) in percentage quantities. Present in much smaller amounts are vitamins (A and B mainly), minerals (including calcium, sodium, potassium, chlorine and phosphorus) and antibodies. Human milk also contains lactoferrin, a protein that inhibits growth of bacteria in the baby's intestine.

mimicry an adaptively evolved resemblance of one species to another species. Mimicry occurs in both the animal and plant kingdoms but is predominantly found in insects. The main types of mimicry are:

(1) Batesian mimicry, named after the British naturalist H. W. Bates (1825–92)—where one harmless species mimics the appearance of another, usually poisonous, species. A good example of this is the

non-poisonous viceroy butterfly mimicking the orange and black colour of the poisonous monarch butterfly. The mimic benefits as, although harmless, any predator learns to avoid it as well as the poisonous species.

(2) Mullerian mimicry, named after the German zoologist J. F. T. Müller (1821–97)—where different species, which are either poisonous or just distasteful to the predator, have evolved to resemble each other. This resemblance ensures that the predator avoids any similar-looking species.

mitochondrion (*plural* **mitochondria**) a double-membrane bound ORGANELLE found in EUCARYOTIC cells. Mitochondria contain circular DNA, RIBOSOMES and numerous ENZYMES, which are specific for essential biochemical processes. Thus, the mitochondrion is the site of such processes as haem synthesis (forms part of HAEMOGLOBIN) and the stages of AEROBIC RESPIRATION that generate most of a cell's ATP. Due to their ATP-producing function, mitochondria are especially abundant in very active, and hence high energy-requiring, cells such as muscle cells or the tails of human sperm cells.

mitosis the process by which a NUCLEUS divides to produce two identical daughter nuclei with the same number of CHROMOSOMES as the parent nucleus. Mitosis occurs in several phases:

(1) Prophase—the condensed chromosomes become visible, and it is apparent that each chromosome consists of two CHROMATIDS joined by a CENTROMERE.

(2) METAPHASE—the nuclear membrane disappears, a spindle forms, and each chromosome becomes attached by its centromere to the equator of the spindle fibres.

(3) Anaphase—each centromere splits, and one chromatid from each pair moves to opposite poles of the spindle.

(4) TELOPHASE—a nuclear membrane forms around each of the group of chromatids (now regarded as chromosomes), and the cytoplasm divides to produce two daughter cells. The stage before and after mitosis is called interphase, and the chromosomes are invisible during this phase as they have decondensed. This is an extremely important part of a cell's cycle as DNA replicates and required proteins are synthesized during interphase.

molality the concentration of a solution expressed in MOLES of solute per one kilogram of solvent. Molality has symbol m and units of mol kg^{-1}.

molar a broad tooth of adult mammals situated at the back of the jaw. Molars are used for grinding food. The third molar (or wisdom tooth) does not appear until early adulthood.

molarity the number of MOLES of a substance dissolved in one litre of solution. Molarity has symbol c and units mol litres^{-1}.

mole the amount of substance that contains the same number of elementary particles as there are atoms in 12 grams of carbon. The number of particles in one mole is called Avogadro's constant and equals 6.023×10^{23} mol^{-1}.

molecular biology the study of the structure and function of large molecules in living cells and the biology of cells and organisms in molecular terms. This applies especially to the structure of PROTEINS and the nucleic acids DNA and RNA.

molecular electronics the use of molecular materials in electronics and opto-electronics. Presently liquid crystalline materials (liquid crystals) are used in displays, organic semiconductors in xerography, and other organic materials are used in devices where the macroscopic properties of the materials are involved. The long term aim of this mixed discipline is to devise molecules or small aggregates of molecules that can exceed the performance of silicon integrated circuits.

molecular formula the chemical formula that indicates both the number and type of any atom present in a molecular substance. For example, the molecular formula for the alcohol ETHANOL is C_2H_6O and indicates that the molecule consists of two carbon atoms, six hydrogen atoms and one oxygen atom.

molecularity the number of particles that are involved in a single step of a reaction mechanism. One step of a mechanism may be termed unimolecular, bimolecular or termolecular, depending on whether there are one, two or three reacting particles. The reacting particles can be ATOMS, IONS or MOLECULES.

molecular weight the total of the atomic weights of all the atoms present in a molecule.

molecule the smallest chemical unit of an element or compound that can exist independently. Any molecule consists of ATOMS bonded together in a fixed ratio, e.g. an oxygen molecule (O_2) has two bonded oxygen atoms and a carbon dioxide molecule (CO_2) has two oxygen atoms bonded to one carbon atom.

mole fraction the ratio of the number of MOLEs of a substance to the total moles present in a mixture.

Molisch's test *see* **alpha-napthol test**.

Mollusca a phylum, with over 50,000 species, that includes the classes Gastropoda, Bivalvia, Cephalopoda (*see individual entries*) and Polyplacophora (the chitons). Most molluscs are marine but there are freshwater and terrestrial varieties. The body takes a similar form throughout the phylum with a *foot* used for movement; a *visceral mass* that includes most of the internal organs; and a *mantle*. The mantle is a fold of tissue over the visceral mass and in many cases it secretes a shell, although not all forms possess a shell. The mantle also houses, posteriorly, a water-filled cavity containing the anus and gills. A RADULA is also a characteristic molluscan feature.

monobasic acid an ACID that contains only one replaceable hydrogen atom per molecule. A monobasic acid will produce a normal SALT during a reaction with a suitable metal.

monoclonal antibody a particular ANTIBODY produced by a cell or cells derived from a single parent cell, i.e. a clone (with each cell being monoclonal). Antibodies thus produced are identical and have specific AMINO ACID sequences. Monoclonal antibodies are used in identifying particular ANTIGENs within a mixture, for example, in identifying blood groups. They are also used to produce highly specific vaccines.

monocotyledon the subclass of flowering plants that have a single seed leaf (cotyledon). Any flowering plant that contains its seeds within an ovary (sometimes forming a fruit) belongs to one of two subclasses, either monocotyledon or dicotyledon (which has two seed leaves). Monocotyledons also differ from dicotyledons in that they have narrow, parallel-veined leaves, their vascular bundles are scattered throughout the stem, their root system is usually fibrous, and their flower parts are arranged in threes or multiples of three. Most monocotyledons are small plants such as tulips, grasses or lilies.

monocyte a large phagocytic cell (*see* PHAGOCYTOSIS), capable of motion, which is present in blood. Monocytes originate in the bone marrow and, after a short residence in the blood, move into the tissues to become MACROPHAGES.

monomer a simple molecule that is the basic unit of polymers. Most monomers contain carbon-carbon double bonds but just have single bonds after they have undergone addition polymerization to form the long-chained polymer. ETHENE is a type of monomer that reacts with other ethene molecules under high pressure and temperatures to form the polymer, polyethene (polythene).

monophyletic refers to any group of organisms that are thought to be derived from the same ancestor.

monopodium the main axis of growth in a plant (e.g. a pine tree) where growth occurs at the tip with branches forming laterally.

monosaccharides (*see also* SACCHARIDES) sugars with the general formula $C_nH_{2n}O_n$ (where n = 5 or 6). They are grouped into hexoses ($C_6H_{12}O_6$) or pentoses ($C_5H_{10}O_5$) on the number of CARBON atoms present and are either aldoses which contain the CHO group or ketoses containing the CO group.

monozygotic twins *see* **identical twins**.

morphine a crystalline ALKALOID occurring in opium. It is used widely as a pain reliever but misuse can be dangerous.

morphogenesis the production of form and structure in an organism through, and differentiation during, its development.

morphology the study of form and structure in organisms, that usually but not exclusively refers to external features. *Anatomy* is the internal aspect of morphology.

motility the ability to move independently.

motor neuron a NEURON that carries impulses from the CENTRAL NERVOUS SYSTEM to an EFFECTOR organ, usually a gland or muscle, producing a response. A *motor unit* is a motor neuron with its nerve terminals and the muscle fibres that undergo innervation. The large muscles of a leg have over two hundred muscle fibres in a motor unit.

mRNA *abbreviation for* MESSENGER RNA.

mucilage a slimy substance (when wet), like a gum, and containing complex carbohydrates. Its function seems to be primarily protective or in the case of plants, as an anchor. Some bacteria are coated with mucilage, to prevent water loss.

mucous membrane tissue that is found as a layer lining cavities in the body that connect with the exterior, e.g. gut, respiratory tracts, etc. It is made up of an epithelium (a protective tissue, of closely packed cells, which has additional functions) which contains mucus-secreting cells and often CILIA.

multicellular a term used to describe tissues, organs or organisms that are made up of a number of cells (*compare* UNICELLULAR) and in which there is a "division of labour" because of differentiation of cells.

multifactorial inheritance is when a particular PHENOTYPE or characteristic is determined by many GENES contributing to the total expression (as with height, for example). Environmental factors also affect the overall outcome, particularly in relation to manifestation of a disease.

muscle a tissue consisting of fibres that can contract to produce movement or tension in the body. There are three types: voluntary or STRIATED MUSCLE, e.g. at joints; involuntary or SMOOTH which is supplied by the AUTONOMIC NERVOUS SYSTEM and effects movements of the intestine, respiratory tract, etc; and CARDIAC MUSCLE, which is found only in the heart.

mutagen any agent that produces an increase in the MUTATION rate of a population. The effect is usually created by chemical alteration of one or more bases or nucleotides in the DNA of the GENES. Chemicals and radiation (e.g. X-rays) are mutagenic.

mutation a change, whether natural or artificial, spontaneous or induced, in the constitution of the DNA in CHROMOSOMES. Mutation is one way in which genetic variation occurs (*see* NATURAL SELECTION) since any change in the GAMETES may produce an inherited change in the characteristics of later generations of the organism. Mutations can be initiated by IONIZING RADIATIONS and certain chemicals.

mutualism a form of SYMBIOSIS in which both participants benefit and in which each must accommodate any changes in the other. There are many examples, including micro-organisms in the guts of termites and ruminants to digest cellulose, making available readily digestible compounds; and NITROGEN FIXATION by bacteria in root nodules.

mycelium a term for the network or mass of fungal HYPHAE that form the body of the fungus, and the vegetative phase.

mycology the study of fungi.

mycoplasmas a group of bacteria within the Eubacteria (the branch of PROCARYOTES in which most modern forms belong, e.g. CYANOBACTERIA) that are believed to be the smallest of all cells (diameter as little as 0.1μm). The cells lack walls (the only procaryote to do so). Some varieties are SAPROTROPHS and others are plant and animal PATHOGENS. There are six genera with about 60 species, and the first was described from a disease in cattle.

myelin sheath *or* **medullary sheath** a layer of fatty material wrapped tightly in a spiral around the AXONS of most vertebrates and some invertebrates. The sheath prevents leakage of current across the axon membrane, thus facilitating rapid transmission of nerve impulses. The layers of membrane forming the sheath are derived from SCHWANN CELLS and the sheath "necks" at intervals along its length, at points called the NODES OF RANVIER.

myeloid tissue the region in BONE MARROW where red blood cells are produced.

myoglobin an oxygen-carrying globular protein comprising a single POLYPEPTIDE chain and a haem group. The haem is an iron-containing molecule consisting essentially of a PORPHYRIN that holds an iron atom as a *chelate* (*see* CHELATION). The iron can reversibly bind oxygen in haemoglobin and myoglobin. The oxygen in myoglobin is released only at low oxygen concentrations, for example, during hard exercise when the muscles demand more oxygen than the blood can supply. Thus myoglobin provides a secondary or emergency store of oxygen. Mammals such as whales have large amounts of myoglobin in their muscles to facilitate diving.

myopia *or* **short-sightedness** occurs when the LENS in the EYE focuses the light in front of the RETINA, usually because the eyeball is too long.

myosin a large protein found originally with ACTIN in muscles but occurring in most EUCARYOTIC cells. There are at least two varieties, one of which is involved in cell locomotion and the other in muscle contraction. The myosin in muscles provides, with actin, the molecular basis of muscle contraction.

Myriapoda a class of segmented terrestrial arthropods (*see* ARTHROPODA) including the centipedes and millipedes (sub-classes Chilopoda and Diplopoda respectively. In some classifications Myriapoda is defunct with Chilopoda and Diplopoda becoming classes). All possess a distinct head with ANTENNAe, MANDIBLEs and a pair of MAXILLAe. Centipedes are carnivorous and bear one pair of legs per segment; millipedes are herbivorous and have two pairs of legs per segment.

N

NAD (*abbreviation for* nicotinamide adenine dinucleotide) is derived from the B vitamin NICOTINIC ACID and is a coenzyme (*see* COFACTOR) important in many biological dehydrogenation reactions, especially in the KREBS CYCLE and ELECTRON TRANSPORT SYSTEM. It is widely found in animal cells. NAD forms a loose bond to its enzyme and is usually positively charged. In the Krebs cycle it accepts one hydrogen atom and two electrons to become a reduced form, NADH. NADH reverts to NAD+, giving up two electrons and one proton to the electron transport system. In the process, and for each molecule of NADH, three molecules of ATP are produced. Another coenzyme, NADP (nicotinamide adenine dinucleotide phosphate) functions in a very similar way and differs only in the possession of an additional phosphate group.

nanometre a unit of measurement that is used for extremely small objects. One nanometre equals one thousand millionth of a metre.

naphtha a mixture of HYDROCARBONS obtained from several sources including coal tar and PETROLEUM. Naphtha from coal tar contains mainly AROMATIC hydrocarbons whereas petroleum naphtha hydrocarbons are predominantly aliphatic.

narcotic any drug that tends to induce unconsciousness or stupor, which is used medically to relieve pain, e.g. opium and morphine.

nares (*singular* **naris**) the paired nostrils present in all vertebrates that are the openings of the nasal cavity. In air-breathing vertebrates, internal nares or choanae are present and these open into the cavity of the mouth. External nares, present in all vertebrates, may occur on a nose, but may be blind-ended cups as in most fishes.

nasal cavity the cavity found in the head of vertebrates which is lined by a mucous membrane containing numerous sensory receptors responsive to smell (OLFACTION). It opens to the outside by means of external NARES on the surface of the head and may be connected to the mouth and respiratory system by means of internal nares.

nastic movement a movement in a plant organ that is triggered by an external environmental stimulus but does not depend upon the direction of that stimulus. These movements are variously described according to the nature of the stimulus involved. *Seismonasty* describes a response to pressure or touch, e.g. the leaves of a *sensitive plant* droop when lightly touched. A rise in temperature causes some flowers, e.g. crocus, to open, and this is known as *thermonasty*. A change in the intensity of light causes other flowers to open, those of the evening primrose at night, for example, and this is called *photonasty*.

natural background the radiation due to natural RADIOACTIVITY and cosmic rays, which must be accounted for in the detection and measurement of radiation.

natural gas the description usually applied to gas associated with PETROLEUM, which originated from organic matter. The gas is a mixture of HYDROCARBONS, mainly methane and ethane with propane. Also present in small amounts may be butane, nitrogen, carbon dioxide and sulphur compounds (*see also* FOSSIL FUELS).

natural selection the process by which evolutionary changes occur in organisms over a long period of time. Darwin explained natural selection (*see* DARWINISM) by arguing that organisms that are well adapted to their environment will survive to produce many offspring, whereas organisms that are not so well adapted to their environment will not. As the better adapted organisms are successfully transmitting their genes from one generation to the next, these are the organisms that are "selected" to survive.

Neanderthal man an early man, *Homo neanderthalensis*, who may have been a subspecies of *Homo sapiens*. Neanderthals occurred about 100,000 years ago and are named from fossil remains first discovered in the Neander valley of West Germany in 1856. Those remains that have been found seem to show that Neanderthals possessed low brows, were quite short but with a brain size similar to that of modern man. They were probably nomadic hunter-gatherers and seem to have been replaced by CROMAGNON MAN.

necrosis death of tissues and cells, while still a part of a living body, generally through disease.

nectar a liquid produced by the nectary (fluid-secreting gland) of flowers that contains sugars and other nutrients. It attracts insects and other animals, which, in obtaining their nectar meal, also pollinate the plant.

nekton any animals that swim in the pelagic zone of the sea or a lake, examples being jelly fish, fish, turtles and whales (*see also* PLANKTON).

nematocyst found only in the CNIDARIA and often located on tentacles, it is a fluid-filled sac containing a long coiled thread, the cnidocil. The whole is contained within a cell called a CNIDOCYTE. The thread is suddenly ejected from the nematocyst in response to touch or a chemical stimulus. It may wrap around and entrap part of a prey organism or puncture it and inject a paralysing poison.

Nematoda the roundworms, which, as the name suggests, have a characteristically rounded body that tapers at either end. They are pseudocoelomates (animals with a COELOM that is not completely lined with mesoderm) and some are very small (about 1 millimetre) while others may reach 1 metre. There are numerous species and they may inhabit water, leaf litter and damp soil and also occur in plant tissues and as PARASITES in animals. They have a tough COLLAGENOUS CUTICLE, which may be transparent, which is shed four times to enable the animal to grow. The cuticle provides a firm surface against which longitudinal muscles contract producing a thrashing motion. Muscles are all longitudinal and this is unique to the roundworms. They have a complete gut with a sucking pharynx and terminal anus. The pseudocoelomic fluid acts as a transport system for nutrients. Sexual reproduction usually occurs and the sexes are generally separate with internal fertilization taking place. The fertilized eggs are typically very resistant and able to exist in harsh conditions. Free-living nematodes are of vital importance as decomposers in the soil and in the recycling of nutrients. Others are grave economic pests attacking the roots of crop plants, etc. Yet others are serious parasites of man and animals; over 50 nematode species can occur in humans including *Trichinella spirallis*, which causes trichinosis.

neo-Darwinism also known as *modern synthesis*, this is the widely held contemporary theory of evolution that was devised during the early decades of the twentieth century. It combines Darwin's theory of evolution by natural selection with modern genetic knowledge about the behaviour of genes and chromosomes and also includes Mendelian heredity (*see also* DARWINISM and MENDEL'S LAWS OF GENETICS).

Neolithic also known as the New Stone Age and describing a period of human history, following on from the PALAEOLITHIC, that commenced about 100,000 years ago. During this period man first developed agriculture and domesticated animals and produced sophisticated stone tools and bone ornaments. It had its origins in the Middle East (Mesopotamia) and radiated from there.

neoplasm an abnormal growth of cells, which may be a benign tumour or, if METASTASIS occurs (dispersal to a new site, usually through the circulation) or if there is invasion of an organ, is generally cancerous.

neotenin *see* **juvenile hormone**.

neotony the situation in which larval traits or juvenile features are retained in an otherwise adult animal. The best known example is the axolotl, which retains the larval gills and is at a juvenile stage although possessing reproductive organs and the ability to breed. If this animal is injected with THYROID HORMONE it metamorphoses, loses its lungs and becomes an air-breathing salamander. Neotony is thought to have been a significant evolutionary mechanism in some animals, including man.

neotype an organism selected to replace the HOLOTYPE if this becomes lost or damaged.

nephridium (*plural* **nephridia**) a tubular structure that is found in some invertebrate animals (e.g. annelids, platyhelminths, rotifers) and opens into the gut or externally via a nephridiopore. Nephridia are formed by a folding in of the ECTODERM and develop independently of the COELOM. *Metanephridia*, present in adult annelids, are nephridia that open into the coelom via openings called nephrostomes. *Protonephridia* are more commonly found in larval animals. They end internally either in solenocytes or flame cells and develop from a hollowing out of nephridial cells. Nephridia transport excretory products that are moved along by CILIA or FLAGELLA. Sometimes they carry gametes, and they may play an important part in osmoregulation.

nephron *see* **kidney**

neritic zone the shallow-water marine zone near the shore, which extends from low tide to a depth of approximately 200 metres. Most BENTHIC organisms live in this zone because sunlight can penetrate to such depths. Sediments deposited here comprise sands and clays with features such as ripple marks.

nerve a long narrow strand of tissue containing NEURONS and GLIAL CELLS, the whole contained within a sheath of connective tissue called the perineurium. A motor nerve carries only MOTOR NEURONS (e.g. to muscles) while a sensory nerve contains sensory neurons (e.g. from sense organs). Often, however, mixed nerves occur carrying both types of neuron. Each neuron acts independently of another within the nerve even though they lie close together. Nerves are the connecting pathways between all the areas and organs of the body and the CENTRAL NERVOUS SYSTEM.

nerve cord a rod containing a bundle of NERVE fibres that runs longitudinally through an animal's body and is an important part of the CENTRAL NERVOUS SYSTEM. It is usually joined anteriorly to the BRAIN, and all members of the CHORDATA have a dorsal hollow nerve cord. In vertebrates, this is the SPINAL CORD. Invertebrates generally have a pair of solid nerve cords running ventrally, with paired segmental ganglia along the length.

nerve impulse *see* **impulse**.

nerve net a network of nerve cells that may be connected to each other either by fusion (asyncitium) or via SYNAPSES. This is commonly found in echinoderms (*see* ECHINODERMATA) and CNIDARIA, where it forms the major part of the nervous system located in the body wall.

nervous system the whole system of tissues and cells (NERVES, NEURONS, GLIAL CELLS, SYNAPSES, RECEPTORS) present in all multicellular animals except sponges, that integrates, co-ordinates and controls all the information related to an animal's internal and external environment. The nervous system operates, i.e. it transmits information, by means of impulses that travel along the neurons (*see* ACTION POTENTIAL). These are conducted rapidly in many different directions and are highly integrated and co-ordinated especially in animals such as the vertebrates. The nervous system enables rapid responses to be made along through-conducting pathways from one site to another. It consists of the CENTRAL NERVOUS SYSTEM (BRAIN and SPINAL CORD in vertebrates and nerve cord and ganglia in invertebrates), and the PERIPHERAL NERVOUS SYSTEM (*see also* AUTONOMIC NERVOUS SYSTEM).

neural arch found in the backbone, an arch of bone that arises from the centrum of each VERTEBRA forming a ring. Thus a tunnel (neural canal) is formed through the vertebral column that houses and protects the SPINAL CORD.

neural tube embryonic nervous tissue, present in chordate animals, that produces the CENTRAL NERVOUS SYSTEM. A *neural plate* of dorsal ECTODERM (just above the NOTOCHORD) rolls up to form the *neural folds*. These then fuse in the mid-dorsal line to form the neural tube, the whole process being known as *neurulation*. If neurulation fails, it results (in humans) in the condition known as spina bifida or even anencephaly (where the skull and cerebral hemispheres do not develop). Anteriorly the neural tube expands to form the brain and it remains narrower posteriorly, becoming the spinal chord.

neurocranium *see* **cranium**.

neuroendocrine system one of a number of dual control systems that regulate certain functions in the body of animals such as the higher vertebrates, and involve both the nervous and ENDOCRINE SYSTEM. Examples are the adrenal medulla (*see* ADRENAL GLAND) and the anterior PITUITARY GLAND, which secrete their HORMONES under the influence of direct nervous stimulation.

neurohormone a HORMONE that is secreted by the nerve endings of specialized nerve cells (neurosecretory cells) rather than by an endocrine gland. Neurohormones are transported within axons (*see* NEURON) and then secreted into the bloodstream or directly to the target tissue or organ. Examples are NORADRENALINE (in vertebrates) and ecdysone, the juvenile hormone associated in insects with ECDYSIS and METAMORPHOSIS.

neuromuscular junction the area of membrane between a muscle cell membrane and a motor neuron forming a SYNAPSE between the two. The motor end plate is the part of the muscle membrane lying beneath the nerve and usually one terminal branch of a neuron forms a synapse with one skeletal muscle fibre. Nerve impulses travel down the neuron and each releases a small amount of ACETYLCHOLINE, which results in a slight depolarization of the endplate. The effects of these are added together (*see* SUMMATION) until a critical threshold level is reached (about -50mV internal negativity). This results in the production of an ACTION POTENTIAL that crosses the synapse into the muscle fibre. This is the mechanism that brings about muscle contraction.

neuron another name for the nerve cell that is necessary for the transmission of information in the form of impulses along its body. The impulses are carried along long, thin structures of the neuron, known as the axon, and are received by shorter, more numerous structures called dendrites. It has been discovered that the transmission of impulses is faster in axons that are surrounded by MYELIN SHEATHS than those that are unmyelinated.

neurotransmitter one of a number of chemical

substances, released in minute amounts by the tip of an AXON, that enables a nerve impulse to cross a SYNAPSE. It is released into the synaptic space and diffuses across to the opposite (postsynaptic) membrane. It may have an excitatory effect, depolarizing the postsynaptic membrane and resulting in the production of an ACTION POTENTIAL. Or it may have an inhibitory, hyperpolarizing effect. Examples of neurotransmitters are, most commonly, ACETYLCHOLINE and also ADRENALINE, NORADRENALINE and dopamine (*see* L-DOPA).

neutral a term indicating that a solution or substance is neither acidic nor alkaline. The most well-known example of a neutral substance is pure water, which should have a pH of seven.

neutralization a reaction that either increases the pH of an acidic solution to neutral seven or decreases the pH of an alkaline solution to seven. Acid-base neutralizations occur in the presence of an INDICATOR, which undergoes a colour change when the reaction is complete.

neutron an uncharged particle that is found in the nucleus of an ATOM. The MASS of the neutron is 1.675×10^{-24}g, which is slightly larger than the mass of the PROTON.

newton the standard international unit of force. One newton (N) is the force that gives one kg an acceleration of one ms^{-2}.

niacin *see* **nicotinic acid**.

niche all the environmental factors that affect an organism and its community. Such factors include spatial, dietary and physical conditions necessary for the survival and reproduction of a SPECIES. Only one species can occupy a specific niche within a community, as the coexistence of any species is subject to distinctions in their ecological niches.

nicotinic acid *or* **niacin** one of the vitamins of the B complex, which is synthesized in animals and plants from the AMINO ACID tryptophan. Tryptophan is found in eggs, milk and cheese, and niacin in yeast, liver, sunflower and groundnut oils. Nicotinic acid deficiency in humans causes the disease pellagra. The amino part of this vitamin, nicotinamide, is a component of the coenzymes NAD and NADP.

nictitating membrane a transparent membranous third eyelid, present in amphibians, reptiles, birds and some mammals. that lies deeper than the other two eyelids. Its (often very rapid) movement is independent of the other eyelids, and it flicks across the cornea, cleaning it and keeping it lubricated.

nidation *see* **implantation**.

nitrification the process by which soil bacteria convert ammonia (NH_3) present in decaying matter into nitrite (NO_2^-) and nitrate (NO_3^-) ions. The produced nitrate ions are taken up by plants and are used in their protein synthesis. Nitrification requires free oxygen, as the bacteria *Nitrosomonas* and *Nitrobacter* oxidize ammonia and nitrites respectively.

nitrogen a colourless, odourless gas, which exists as N_2 and forms almost 80 per cent of the atmosphere by volume. It is obtained commercially by fractionation of liquid air and is used extensively in producing AMMONIA, as a refrigerant, and as an inert atmosphere. It is vital for living organisms occurring in PROTEINS (and NUCLEIC ACIDS), and the NITROGEN CYCLE is the circulation of the element through plant and animal matter into and from the atmosphere.

nitrogenase the enzyme that catalyses the production of nitrogenous compounds from free NITROGEN in the process of NITROGEN FIXATION.

nitrogen cycle the regular circulation of nitrogen owing to the activity of organisms. Nitrogen is found in all living organisms and forms about 80 per cent of the atmosphere (this proportion is maintained by the nitrogen cycle). The start of the nitrogen cycle can be regarded as the uptake of free nitrogen in the atmosphere by bacteria (NITROGEN FIXATION) and the uptake of nitrate (NO_3^-) ions by plants. The nitrogen is incorporated into plant tissue, which in turn is eaten by animals. The nitrogen is returned to the soil by the decomposition of dead plants and animals. NITRIFICATION converts the decomposing matter into nitrate ions suitable for uptake by plants.

nitrogen fixation the process by which free nitrogen (N_2) is extracted from the atmosphere by certain bacteria. Some free-living bacteria can use the nitrogen to form their AMINO ACIDS, while other nitrogen-fixing bacteria live in the root nodules of leguminous plants (peas and beans) and provide the plants with nitrogenous products. This enables the plant to survive in nitrogen-poor conditions while the bacteria has access to a carbohydrate supply in the plant (a symbiotic relationship). The nitrogen-fixing bacteria are able to convert free nitrogen into nitrogenous products because of the presence of the enzyme nitrogenase within their cells.

noble gases the elements comprising group 8 of the PERIODIC TABLE. Noble gases are usually referred to as inert gases because of their relative unreactivity.

node the site on a plant stem where the bud and leaves arise.

nodes of Ranvier interruptions along the length of a nerve AXON where there is an absence of MYELIN SHEATH between SCHWANN CELLS. At the nodes of Ranvier the current can flow through the membrane of the axon whereas elsewhere this is prevented by the myelin sheath. The effect is to make impulse transmission even more rapid and it may leap from one node to the next.

nodule a small swelling or structure on a plant, especially the root, which is due to nitrogen-fixing bacteria.

nondisjunction the failure of chromosomal pairs or sister CHROMATIDS to separate during MITOSIS or MEIOSIS respectively. Nondisjunction produces daughter cells containing an unequal number of CHROMOSOMES, either too many or too few.

noradrenaline a NEUROTRANSMITTER of the SYMPATHETIC NERVOUS SYSTEM that generally has an excitatory effect. It is secreted by nerve endings and also by the ADRENAL GLANDS, but less commonly than ADRENALINE. It is similar to adrenaline both in structure, being derived from the amino acid tyrosine, and also in its general function. However, it is more concerned with maintaining arousal in the brain, with mood and ordinary activity than with the "fright, flight or fight" response. It is also known as norepinephrine.

normal salt any SALT, formed by an ACID, that loses more than one hydrogen ion (H^+) per molecule during a NEUTRALIZATION reaction. It should be noted, however, that the production of a normal salt is dependent on the quantities of the acid and BASE used in the reaction.

notochord a skeletal rod of tissue enclosed by a firm

sheath, which is present at some stage in all CHOR-
DATES. However, in vertebrates it occurs as a complete entity only in the embryo but may persist very
much reduced in adults, between the vertebrae that
replace it. It is situated above the ALIMENTARY CANAL
and beneath the nerve cord and provides a protago-
nist for muscles as well as for support.

nucellus the tissue surrounding the OVULE in a
flowering plant or seed of a GYMNOSPERMAE (e.g. a
conifer), providing nutrition.

nuclear magnetic resonance spectroscopy an
analytical technique based upon the absorption of
electromagnetic radiation in the radio frequency
region for those nuclei which spin about their own
axes. The result is a change in the orientation of the
spinning nuclei in a magnetic field. Of all the susceptible nuclei (^{1}H, ^{19}F, ^{31}P), ^{1}H is the most sensitive.
The technique is used for the identification of or-
ganic materials and magnetic resonance has been
developed to form a medical imaging tool.

nuclear membrane a double membrane that surrounds the nucleus of EUCARYOTIC cells. There is a
space between the outer and inner nuclear membranes, but both membranes seem to fuse together
at regions containing nuclear pores. Nuclear pores
contain proteins that probably control the exchange
of material between the cell nucleus and its CYTO-
PLASM.

nucleic acid a linear MOLECULE that acts as the
genetic information store of all cells. Nucleic acids
occur in two forms, deoxyribonucleic acid (DNA)
and ribonucleic acid (RNA), but both forms are
composed of four different NUCLEOTIDES, which react
to form the long chain-like molecule. DNA is found

inside the nucleus of all EUCARYOTES as it is the major part of CHROMOSOMES, but RNA is found outside the nucleus and is essential for TRANSCRIPTION and TRANSLATION during protein synthesis.

nucleolus a membrane-bound object found within the NUCLEUS. The nucleolus contains gene sequences that code for ribosomal RNA, ribosomal RNA itself, and proteins necessary for rRNA synthesis.

nucleoprotein present in the cells of living organisms and consisting of nucleic acid (DNA, RNA) and proteins associated in a complex. This forms the CHROMOSOME and the protein component is mainly HISTONE.

nucleoside an organic compound that consists of a RIBOSE or deoxyribose sugar linked to a purine or pyrimidine base. The ribose forms occur in RNA and deoxy forms in DNA.

nucleotide a MOLECULE that acts as the basic building block of the NUCLEIC ACIDS, DNA and RNA. The structure of a nucleotide can be divided into three parts—a five-carbon sugar molecule; a phosphate group; and an organic base. The organic base can be either a PURINE, e.g. ADENINE, or a PYRIMIDINE, e.g cytosine.

nucleus (*plural* **nuclei**) in biology, the large ORGANELLE found in the CYTOPLASM of all plant and animal cells (EUCARYOTES), but not bacteria, and which contains the genetic material, the DNA. It has a double membrane, punctuated by numerous nuclear pores. These allow for traffic of material between the cell's cytoplasm and its nucleus. Unless the cell is actively dividing, a NUCLEOLUS is present within the nucleus, which consists of DNA, RIBOSOMAL RNA and PROTEIN. The nucleus is filled with a thick fluid sap called nucleoplasm and in the non-dividing cell, the genetic material is dispersed within it and is known as chromatin. The nucleolus disappears when the cell divides and the chromatin becomes organized as CHROMOSOMES. The nucleus is the cell's vital control centre.

In chemistry and physics, the term nucleus refers to the small, positively charged core of an ATOM that contains the PROTONS and NEUTRONS. The electrons of an atom orbit the nucleus.

null hypothesis a hypothesis that examines the existence of a specific relationship by enabling it to be statistically tested. A null hypothesis assumes that the expected results from an experiment have just arisen from chance with no significant change occurring. These expected results are statistically compared with the observed results derived from the experiment. Any significant difference between the expected and the observed results will be taken as evidence to support the experimental hypothesis and to rule out the null hypothesis.

O

occipital condyle a bony projection arising from the occipital bone at the back of the skull and articulating with the first cervical vertebra of the spinal column, the *atlas*. There are a pair of occipital condyles in amphibians and mammals and a single one in birds and reptiles. They are absent in most fish, which are unable to move their heads.

oceanic zone the open deep sea environment beyond the continental shelf, where the depth is in excess of 200 metres.

oceanography the study of the oceans, including tides, currents, the water and the sea floors.

ocellus (*plural* **ocelli**) one of a number of different kinds of simple eye found in the invertebrates. It occurs in insects, flatworms and other arthropods (*see* ARTHROPODA) and usually consists of cells sensitive to light and a single lens.

Odonata an order of the class INSECTA comprising the damselflies and dragonflies. They are carnivorous both as adults and nymphs, and most are tropical species. Eggs are laid in or near water, and aquatic nymphs hatch out, which have gills for respiration. They are voracious hunters with biting mouthparts that are borne on a mask held beneath the head. This is shot out suddenly to seize the prey. The nymph crawls up a stalk out of the water prior to its final moult. The adults have a pair of large COMPOUND EYES and two pairs of conspicuously veined membranous wings. In dragonflies, the wings are held out horizontally at right angles to the body at rest, whereas in damselflies they are folded over the back. The legs catch insect prey in flight and there is a long slender abdomen. They are often brightly coloured and are an ancient group; some fossil forms found in CARBONIFEROUS rocks had wingspans of over half a metre.

odontoblast a cell found in the lining of the PULP CAVITY of a vertebrate tooth. Odontoblasts bear processes that extend into the DENTINE (the main calcareous substance of which a tooth is composed), which they produce.

oedema an abnormal swelling of tissue due to an increase in fluid volume either in tissue spaces or sacs of the body, e.g. pulmonary oedema.

oesophagus the first part of the ALIMENTARY CANAL, lying between the PHARYNX and the stomach in animals, which may contain a CROP. Its function is to pass food to the stomach, achieved by muscular movement known as PERISTALSIS.

oestrogens a collective term for a group of STEROID hormones found in female mammals. Oestrogens are secreted mainly by the ovaries and to a small extent by the adrenal cortex. They are also produced in small animals by the testes of male mammals. They are responsible for the development of the secondary sexual characteristics in human females (e.g. growth of breasts, changes in the profile of the pelvic girdle, growth of pubic hair) and deposition of body fat. Oestrogens regulate the OESTROUS CYCLE in other mammals. They are produced at high levels by the GRAAFIAN FOLLICLES of the ovary at the time of ovulation, preparing the uterus for pregnancy (*see* MENSTRUAL CYCLE). Oestrogen and PROGESTERONE act together in regulating the female reproductive cycle. Naturally occurring oestrogens in humans include oestradiol, which is the most important, and also oestriol and oestrone. They are used medically, together with progesterone, in the contraceptive pill and also to treat other gynaecological disorders in women.

oestrous cycle the reproductive cycle present in most mature non-pregnant female mammals other than primates that lasts between 5 and 60 days depending upon the species. There are several phases in the cycle the first being the *pro-oestrus*, when the GRAAFIAN FOLLICLES mature and produce OESTROGENS. This is followed by the *oestrus*, or heat, when the female is sexually attractive to a male and is ready to mate. *Metoestrus* is the stage when the CORPUS LUTEUM develops, and *dioestrus* is when this secretes PROGESTERONE in order to prepare the uterus for the implantation of an embryo. Males may have a cycle of sexual activity corresponding to that of the females. Some mammals are *polyoestrous* and have numerous cycles per year, and the males may then be sexually active all the time. In other mammals there is a well-defined breeding season, often triggered by environmental factors, and only one oestrous cycle. These are described as being *monoestrous*.

oil a greasy liquid substance obtained from animal or vegetable matter (*see* FATS) or from mineral matter (*see* PETROLEUM).There are essentially three groups of oils. The fixed or fatty oils from animal, vegetable and marine sources occur as glycerides and ESTERS. Mineral oils are HYDROCARBONS and are produced from petroleum, shale and coal, and essential oils are derived from certain plants and tend to be volatile hydrocarbons.

oleic acid a widely occurring, unsaturated FATTY ACID that has one double bond and is a constituent of many plant and animal fats. It is found, for example, in soya bean oil, groundnut oil and butterfat.

olfaction the process by which an animal senses smells. Olfactory organs contain sensory receptors that are sensitive to chemicals borne in the air or water. They are found, for example, in the nose of vertebrates. These receptors transfer information via sensory NEURONS to an area in the brain known as the olfactory bulb. The olfactory nerve transmits information from the olfactory bulb to areas within the CEREBRAL CORTEX where it is interpreted.

Oligocene a geological epoch that occurred 38 to 24.6 million years ago and is the third part of the TERTIARY period. It was followed by the MIOCENE and during this time mammals continued to flourish and diversify.

Oligochaeta a class of the phylum ANNELIDA that are mostly terrestrial or freshwater organisms. The most familiar is the earthworm, *Lumbricus*. Oligochaetes have a reduced head without appendages, and they have few CHAETAE, which are not borne on parapodia (*compare* POLYCHAETA). They are HERMAPHRODITE worms with GONADS restricted to a few segments and with internal fertilization There are no larvae, and development is direct.

oligotrophic a term that describes a lake or river with a poor supply of nutrients, hence there is a low rate of formation of organic matter. There are few plants or animals.

ommatidium *see* **compound eye**

omnivore any organism that eats both plant and animal tissue.

oncogene any gene directly involved in cancer. Oncogenes may be part of a specific VIRUS that has managed to penetrate and replicate within the host's cell, or they may be part of the individual's GENOME, which has been transformed by radiation or a chemical.

oncogenic any factor that causes cancer whether it be an organism, a chemical substance or something occurring in the environment. Some viruses are oncogenic to vertebrate animals and may contain ONCOGENES, which make a normal cell become a cancer cell.

ontogeny the complete development of an individual to maturity.

oocyte a cell that undergoes MEIOSIS to form the female reproductive cell (egg or ovum) of an organism. In humans, a newborn female already has primary oocytes, which will undergo further development when puberty is reached but which will only complete secondary meiosis to form the secondary HAPLOID oocyte if fertilization occurs.

oogenesis the production and growth of eggs (ova) within the ovary of a female animal. Cells called oogonia divide numerous times by MITOSIS to become OOCYTES, which are the prospective eggs. These undergo various stages of MEIOSIS which, in human females, is arrested after birth and only continues at puberty. Eventually a secondary oocyte is produced, which itself only completes meiosis when fertilized, thus also finishing oogenesis. Hence, while all the potential egg cells are present at birth, oogenesis is a prolonged process that only a few will complete.

ooze a deep sea mud made up of clays and the calcareous or siliceous remains of certain organisms, e.g. diatoms.

open chain a compound with an open chain not a ring structure, as in aliphatic compounds, e.g. ALKANES, ALKENES and ALKYNES and their derivatives.

operculum (*plural* **opercula**) a lid or flap. In animals it covers an opening where it occurs, and an example is the calcareous plate with which a gastropod mollusc closes the aperture of its shell. An operculum covers the GILL SLITS of larval amphibians and fish. In plants, it is present as a conical cap opening the SPORANGIUM of mosses, and is also part of the cell wall in some fungi.

opiate one of a number of drugs derived from opium, which itself comes from the poppy plant. Opiates have a NARCOTIC effect depressing brain function and are used medically in pain control. Morphine is an opiate as is heroin (*see* ENDORPHIN).

optical activity *or* **optical rotation** is when a solution of a substance rotates the plane of transmitted polarized light. It occurs with optical isomers, with one form rotating the light in one direction and the other rotating the light by the same amount in the opposite direction (DEXTROROTATORY and LAEVOROTATORY COMPOUNDS). A racemic mixture is optically inactive because it contains equal amounts of both forms.

optical isomerism the existence of two chemical compounds that are isomers, which form non-superimposable mirror images.

optical rotation *see* **optical activity**.

optic chiasma a mass of nerve fibres occurring beneath the FOREBRAIN in vertebrates. It is formed from nerve fibres of the left optic nerve crossing to the right hand side of the brain and those of the right optic nerve crossing to the left. All the fibres cross over on most vertebrates but in mammals about half of them do so.

optic nerve present in vertebrates, this is the second CRANIAL NERVE, which is a paired sensory nerve carrying information to the brain from each EYE (*see also* RETINA).

orbit a more or less rounded cavity in the vertebrate skull, the eye socket, that contains the eyeball.

orbitals orbitals are a means of expressing, rationalizing and correlating atomic structure, bonding and similar phenomena. The Bohr theory postulated the positioning of ELECTRONS in definite orbits about a central NUCLEUS. However, it was soon discovered that this was too simple and that electrons behave in some ways as waves, which makes their spatial position more imprecise. Hence the old "particle in an orbit" picture was replaced by an electron "smeared out" into a charge cloud or orbital, which represents the probability distribution of the electron. An atomic orbital is thus one associated with an atomic nucleus and has a shape determined by QUANTUM NUMBERS. Various types of orbital, designated s, p, d, etc, are distinguished. An s orbital is spherical, and a p orbital is dumbbell-shaped. When two atoms form a COVALENT BOND, a molecular orbital with two electrons is formed, associated with both nuclei (*see* σ and π BOND). The overlapping of atomic orbitals in a carbon-carbon single bond (e.g. ETHANE) creates a molecular orbital centred on the line joining the two nuclei. In a carbon-carbon double bond (e.g. ETHENE), the second of the two bonds is created by two overlapping p orbitals, forming the π bond. The two overlapping dumbbells create two sausage-like spaces of electron "cloud" on each side of the line joining the nuclei. BENZENE has torus-shaped (doughnut-shaped) molecular orbitals on each side of the ring because of overlap and merging of p atomic orbitals.

order a taxonomic grouping used in the classification of organisms, which may contain one or more families. Similar orders are themselves grouped together within a class.

order of magnitude the approximate size of an object or quantity usually expressed in powers of 10.

ordinal number in general, 1st, 2nd, etc, rather than 1, 2, etc (cardinal numbers).

ordinal scale a statistical scale that arranges the data in order of rank in the absence of a numerical scale with regular intervals. Ordinal scales are ideal for data that contain relationships such as bad, good, better, best, as the data can be put in rank order but no regular interval can be measured between the ranked judgements.

ordinate the vertical or y-axis in a geometrical diagram for Cartesian co-ordinates. For example, a point with co-ordinates (2,-6) has an ordinate of -6.

Ordovician the second geological period of the Palaeozoic era, which lasted from 505 to 438 million years ago. It followed the CAMBRIAN and preceded the SILURIAN and was named by a British geologist, Charles Lapworth (1842-1920). Ordovician rocks laid down in deep sea environments contain important fossils, including GRAPTOLITES, which are the most abundant type. Other fossils include BRYOZOANS, BRACHIOPODS, TRILOBITES, echinoids (*see* ECHINODERMATA), nautiloid CEPHALOPODS and crinoids. CORALS first appeared during the Ordovician.

organ any distinct and recognizable unit in terms of its structure and function that occurs within an organism. It contains two or more types of tissue and examples in plants are stems, leaves, roots and flowers and in animals, the kidneys, liver, skin, ears and eyes are all organs.

organelle any functional entity that is bound by a membrane to separate it from the other cell constituents. Organelles are found in the cells of all EUCARYOTES and include the CHLOROPLASTS and vacuoles of plant cells in addition to the NUCLEUS, MITOCHONDRIA, GOLGI APPARATUS, ENDOPLASMIC RETICULUM, and other small vesicles of both animal and plant cells.

organic chemistry the branch of chemistry concerned with the study of carbon compounds. Organic chemistry studies the typical bond arrangements and properties of carbon compounds containing hydrogen and, less frequently, oxygen and nitrogen. As most organic compounds are derived from living organisms, two major areas for study are the biologically important organic compounds and the commercially important organic compounds, e.g. ALKANES derived from oil.

ornithine *see* **urea cycle**

Orthoptera a large order of the class INSECTA that contains the grasshoppers, crickets, locusts, preying mantis and cockroaches. They are usually quite large and herbivorous with biting mouth parts. Two pairs of wings may be present but many are wingless. The forewings are leathery and narrow and the hind wings are membranous and folded under them. The hind legs are often large and may be modified for jumping. Many species produce sound by STRIDULATION. The stridulatory organs often involve the wings that are rubbed together or part of the hind leg that is rubbed against the forewing. Some species such as locusts and cockroaches are pests of crops or in buildings.

orthotropism a growth response in a plant (TROPISM) that tends to be orientated with that of the stimulus direction. An example is the vertical growth of main stems and roots that respond directly to gravity (orthogeotropism, *see also* PLAGIOTROPISM).

osmoregulation any process or mechanism in animals that regulates the concentration of salts and water in its body. There is a tendency for water to pass into or out of an animal's body by OSMOSIS, depending upon the environment it inhabits. Water tends to pass into the body of a freshwater animal as the concentration of salts in its body is higher than in the external environment. Various methods are used to rid the body of excess water, such as CONTRACTILE VACUOLES in simple organisms and kidneys in fish. In marine animals there is a tendency to lose water, and the kidneys are the main osmoregulatory organs. In terrestrial animals the outer skin or cuticle helps to prevent desiccation or excessive water uptake, but a wide variety of mechanisms operate, involving physical and behavioural adaptations, so that a correct salt and water balance is maintained (*see* KIDNEY and OSMOSIS).

osmosis the process in which solvent molecules (usually water) move through a semi-permeable membrane to the more concentrated solution. Many mechanisms have evolved to prevent the death of animal cells either by too much water entering a cell by osmosis, causing it to rupture, or by too much leaving by osmosis, causing it to shrink (plasmolysis). Such mechanisms include the presence of a pump within the membrane of animal cells, which actively regulates the concentration of vital cellular IONS and the excretion of salt through the gills of marine bony fish to remove the salt gained by diffusion and drinking.

ossification the formation of bone brought about by OSTEOBLASTS, which are specialized cells. *Membrane bone* such as that in the face and skull is formed directly by the osteoblasts within connective tissue. Cartilage bone is formed by the replacement of cartilage of an embryo skeleton, again brought about by the oestoblasts (*see* BONE).

Osteichthyes *see* **bony fishes**

osteoblast a specialized cell that is responsible for the formation of BONE.

osteoclast a specialized cell with many nuclei that breaks down the calcified matrix of bone, a process that accompanies bone growth.

Ostracoda a subclass of the class CRUSTACEA that contains organisms that are mostly small (only a few millimetres long) and inhabit marine and freshwater environments. They have a bivalved carapace that encloses the body completely and is operated by adductor muscles. They are filter feeders with head appendages for collecting food. Small appendages may occur on the body, but they are few in number.

ovary (*plural* **ovaries**) the reproductive organ of female animals that produces ova (eggs) and is usually paired in vertebrates. In fish the ovaries may be fused to form one structure, and birds have only one functional ovary, which is the left one. STEROID HORMONES are produced by the ovary, namely OESTROGEN and PROGESTERONE. The ovary contains numerous GRAAFIAN FOLLICLES, in which the eggs develop. An open-ended tube (OVIDUCT, FALLOPIAN TUBE) is situated close to the ovary, which receives the eggs and leads down into the UTERUS (womb) (*see* MENSTRUAL CYCLE, OESTROUS CYCLE, OOGENESIS and OVULATION). In plants, the ovary is the hollow basal portion of a CARPEL of a flower and contains one or more ovules. A compound ovary consists of two or more fused carpels themselves containing united ovaries. Once fertilization has taken place the ovary wall becomes the fruit that encloses the seeds.

oviduct a tube that carries an egg away from the ovary of an animal sometimes to the outside or to another part of the reproductive system. Muscle action or CILIA operate to transport the eggs along the oviduct. In humans it is called the FALLOPIAN TUBE.

oviparous a term describing animal reproduction where the development of the embryo and subsequent hatching occurs outside the female's body. Oviparous reproduction is found in birds, most fish, and reptiles.

ovoviviparous the term to describe the development of offspring within the body of the female in the absence of a placenta. Ovoviviparous reproduction is found in certain species of fish, reptiles and insects, where the young are retained within the body of the mother solely for protection as they receive their nutrients from the egg and not from the mother.

ovulation the release of an egg from the ovary. In mammals this is brought about by LUTEINIZING HORMONE, which is secreted by the anterior PITUITARY GLAND. A follicle migrates to the surface of the ovary, and when it is ripe it ruptures and releases the egg, which then enters the OVIDUCT or FALLOPIAN TUBE (*see* GRAAFIAN FOLLICLE and MENSTRUAL CYCLE).

ovule a reproductive structure, present in seed plants, that develops into a seed after fertilization. It consists of the integuments (a single or double protective layer), which ultimately form the seed coat. These surround the NUCELLUS, which is a mass of tissue. Each ovule has a stalk attaching it to the carpel. In GYMNOSPERMS the ovules are unprotected and occur on specialized leaves (the ovuliferous scales) of the female cones. In ANGIOSPERMS the ovules are protected within carpels.

ovum (*plural* **ova**) produced by the OVARY, the mature unfertilized, reproductive cell of female animals. It is a spherical cell with an outer membrane and a single nucleus and is non-motile. It is an oocyte in many animals, and may not complete its development until fertilization (*see* OOGENESIS). The equivalent structure in plants is the oosphere.

oxalic acid *or* **ethaneidoic acid** a crystalline acidic substance that is highly toxic and is found in some plants, e.g. rhubarb leaves.

oxidation any chemical reaction that is characterized by the gain of OXYGEN or the loss of ELECTRONS from the reactant. Oxidation can occur in the absence of oxygen, as a molecule is also said to be oxidized if it loses a hydrogen atom.

oxidative phosphorylation *see* **phosphorylation**

oxide a compound formed by the combination of OXYGEN with other elements, with the exception of the inert gases (*see* NOBLE GASES).

oxidizing agent any substance that will gain ELECTRONS during a chemical reaction. Oxidizing agents will readily cause the OXIDATION of other atoms, molecules or compounds, depending on the strength of the oxidizing agent and the reactivity of the other substance. The following are all examples of oxidizing agents arranged in order of increasing oxidizing strength—sodium ions (Na^+), sulphate ions (SO_4^{2-}), and oxygen molecules (O_2).

oxygen a colourless and odourless gas, which is essential for the respiration of most life forms. It is the most abundant of all the elements, forming 20 per cent by volume of the atmosphere; about 90 per cent by weight of water; and 50 per cent by weight of rocks in the crust. It is manufactured by the fractional distillation of liquid air and is used for welding, anaesthesia and rocket fuels.

ozone a denser form of oxygen that exists as three atoms per molecule (O_3). Ozone is a more reactive gas than the more common diatomic molecule (O_2), and can react with some hydrocarbons in the presence of sunlight to produce toxic substances that are irritants to the eyes, skin and lungs. Minute quantities of O_3 are found in sea water. It forms the Earth's OZONE LAYER, 15 to 30 kilometres above the Earth's surface.

ozone layer a region of the Earth's atmosphere

containing ozone that acts as a barrier against the ULTRAVIOLET RADIATION from the Sun. Scientists and environmentalists have announced that large holes are appearing in the ozone layer as a result of the widespread use of the ozone-depleting chemicals called chlorofluorocarbons (CFCs). CFCs are used in many industrial processes and, because of their unreactive qualities, as aerosol propellants. In the Earth's atmosphere, however, they will readily react with, and destroy, ozone in the presence of sunlight. The depletion of the ozone layer is cause for concern as increased exposure to ultraviolet radiation will increase the incidence of skin cancers and eye cataracts. To prevent further damage to the ozone layer, industrial nations are being called upon to greatly reduce the use of CFCs.

P

pacemaker an area of modified and specialized cardiac muscle cells in the mammalian heart. It is located in the right atrial wall near the opening of the VENA CAVA and is responsible for initiating and maintaining the beat of the heart. The cells of the pacemaker are controlled by the AUTONOMIC NERVOUS SYSTEM, producing contractions that stimulate the rest of the heart muscle to contract.

Also, a device that can be implanted into the chest to maintain the heart beat in the stead of a diseased or defective pacemaker.

paedogenesis when reproduction occurs albeit that the animal is in a larval form—a form of NEOTONY. The best known example is the axolotl—*see under* NEOTONY.

palaeobotany the study of fossil plants.

Palaeocene the first epoch of the TERTIARY, which began about 65 million years ago. Due to some cataclysmic event (*see* DINOSAUR), huge numbers of flora and fauna disappeared at the end of the Cretaceous and this led to the ascendancy of the mammals, and by the end of the Palaeocene, primates and rodents had evolved.

palaeoclimatology the study of past climates, based

upon information from sediments deposited and the flora and fauna (indicated by fossils).

palaeoecology the study of fossils to recreate the inter-relationships between the organisms and their environment, utilizing additional information gleaned from the sediments deposited at the time and associated studies of the palaeoclimatology.

Palaeolithic The Old Stone Age, which lasted from approximately two million years to 10,000 years ago. During this period man used stone tools, albeit primitive versions, made from chipped stones and flints.

palaeontology the scientific study of FOSSILS.

palate in vertebrates, the roof of the mouth, which separates the nasal cavity from the buccal (mouth) cavity. The separation of the buccal cavity from the airway permits simultaneous breathing and eating.

pallium *see* **cerebral cortex**.

palp a sensory organ of invertebrates, usually located near the mouth. In polychaete worms (ANNELIDA) there are tactile organs on the head, while in various other organisms the organ serves an olfactory or feeding role.

palynology the study of (mainly fossil) pollen, spores and some other microfossils. It deals primarily with structure, classification and distribution and is useful in petroleum exploration, and palaeoclimatology. In the main, the spores and pollen are highly resistant and in some sedimentary sequences are the only fossils that can be used for stratigraphic correlation.

pancreas a gland in vertebrates that performs both endocrine and exocrine functions. It is between the duodenum and spleen and secretes digestive enzymes, e.g. LIPASE, AMYLASE and trypsinogen (which becomes TRYPSIN). Its endocrine functions involve secretion of the HORMONES INSULIN and GLUCAGON from the ISLETS OF LANGERHANS. Both hormones are active in the control of blood sugar levels (*see also* DIABETES).

Pangaea (*see also* CONTINENTAL DRIFT) the single continent, proposed by Alfred Wegener in 1915, which came into being late in the Permian and lasted into the Triassic. The construction of this "supercontinent" was supported by the geometric fit of coastlines on either side of the Atlantic, the similarity between faunal distributions, and other factors.

pantothenic acid a vitamin in the VITAMIN B COMPLEX and a component of coenzyme A, which is important in the oxidation of fats and carbohydrates. This vitamin occurs in many different types of food, e.g. cereal grains, egg yolk, liver, peas, yeast, and so is rarely a factor in dietary vitamin deficiencies.

papilla (*plural* papillae) a projection from various tissues or organs, mainly in animals. Papillae protrude from the surface of, for example, the tongue to increase the surface area for taste buds.

pappus a ring of fine, feathery hairs, formed from modified sepals at the top of fruits of certain plants. The dandelion is an example where a pappus of hairs rings the top of the thin stalk to form a "parachute" to facilitate dispersal in the wind.

paraffins (*see* ALKANE) the general or informal term for saturated ALIPHATIC HYDROCARBONS with the formula C_nH_{2n+2}. They are quite unreactive, hence the name paraffin (from the Latin *parum affinis*, "little allied").

parasite any organism that obtains its nutrients by

living in or on the body of another organism (its host). Parasites can be either completely dependent on their host for survival (obligate) or are able to survive without their host (facultative). The extent of the damage on the health of the host by parasitic infestation can range from being virtually harmless to so severe that it causes the death of the host. Highly evolved parasites are so well adapted, however, that the host is able to survive and reproduce as normal, thus providing the parasite with a permanent supply of nutrients.

parasympathetic nervous system a part of the AUTONOMIC NERVOUS SYSTEM that often acts antagonistically with the SYMPATHETIC NERVOUS SYSTEM. The two parts differ in most respects, and the parasympathetic nerves originate at the top and bottom of the CENTRAL NERVOUS SYSTEM in the cranial and sacral regions. The AXONS of this system tend to be longer than those of sympathetic nerves, and synapses with other neurons are near to the target organs, and ACETYLCHOLINE is the NEUROTRANSMITTER released. The parasympathetic system contracts the bladder, stimulates the sex organs, decreases the heart rate, promotes digestion, and stimulates salivation, etc.

parathyroid glands paired endocrine glands of higher vertebrates located near or within the THYROID GLAND. The glands produce *parathyroid hormone* (PTH or parathormone). PTH is a polypeptide that controls the levels of calcium in the blood, working in concert with VITAMIN D and calcitonin (a hormone involved in calcium and phosphate levels and their uptake). A low calcium level stimulates hormone secretion, increasing transfer of calcium to the blood from the bones.

parenchyma plant tissue with thin-walled, unspe-
cialized cells with intercellular air spaces. These
cells undertake most of the plant's metabolism,
producing and storing organic compounds. In addition, parenchyma cells can also differentiate into
other types of plant cells.

parthenocarpy is when fruits form without fertilization first. The fruits so formed are seedless, e.g.
banana. It seems plant GROWTH SUBSTANCES may be
involved, and it can be induced in tomatoes by
AUXINS.

parthenogenesis the development of a new individual from an unfertilized egg. Parthenogenesis is
most common among the lower invertebrates, such
as insects and flatworms. For example, the process
can be part of the honey bee life cycle if the HAPLOID
eggs laid by the queen remain unfertilized by sperm.
The larvae from these eggs will develop into the
male bees (drones), whose only function is to produce sperm. If the eggs laid by the queen are fertilized then the larvae develop into sterile female
worker bees or fertile queens, depending on the food
supply. Parthenogenesis also occurs in some plants,
such as the common dandelion, and it can be induced artificially in many species by stimulation of
the egg cell.

parturition the process of giving birth at the end of
pregnancy. The process involves PROSTAGLANDINS,
oxytocin (a peptide hormone) secreted by the PITUI-
TARY GLAND and strong uterine contractions.

passive immunity the ability of an individual to
resist disease using ANTIBODIES that have been donated by another individual rather than by producing its own antibodies. Passive immunity is ob-

tained by young mammals from their mother's milk during the first few weeks of life as the newly born are virtually incapable of antibody production. In humans, breast-fed infants will receive most of their maternal antibodies from their mother's milk, but one antibody, IgG, will be found in all infants, whether breast-fed or bottle-fed, as IgG can cross the placenta during foetal development.

Pasteur, Louis (1822–1895) French chemist and bacteriologist who was the first to demonstrate that a colony could be grown in a culture medium that had been infected with a few cells of the microorganism. This experiment showed that living cells had an inheritance of their own and helped discredit the theory of SPONTANEOUS GENERATION of life. Although Pasteur had been aware of the role of microorganisms in FERMENTATION since 1858, he did not accept their role in causing disease until several years later. He demonstrated that attenuated forms of micro-organisms could be used in innoculation, providing immunization for the host, and in 1885 he produced the first rabies vaccine.

pasteurization a process developed by PASTEUR of partially sterilizing food by heating it to a certain temperature. Food is pasteurized before distribution as the process can destroy potentially harmful bacteria, e.g. heating milk for 30 minutes at 62°C destroys the bacteria responsible for tuberculosis, and increases the shelf life of food by delaying its FERMENTATION.

patella the kneecap, a bone present in most mammals to protect the knee joint. It is a small rounded bone in the tendon in front of the knee.

pathogen any organism that causes disease in another organism. Most pathogens that affect humans and other animals are bacteria or viruses, but in plants there is also a wide range of fungi that act as pathogens.

peat an organic deposit formed from compacted dead and possibly altered, vegetation. It is formed from vegetation in swampy hollows and occurs when decomposition is slow due to the ANAEROBIC conditions in the waterlogged hollow. Sphagnum moss is among the principal source-plants for peat. As peat builds up each year, water is squeezed out of the lower layers causing the peat to shrink and consolidate. Even so, cut peat has a high moisture content and is dried in air before burning. In Ireland and Sweden, peat is used in power stations.

peck order *see* **dominance**

pectic compounds POLYSACCHARIDES composed mainly of sugar acids found in plant cell walls and the middle LAMELLAE betwen the walls of adjacent cells. The compounds (pectic acid, pectates, pectose and PECTIN) are normally insoluble but in ripening fruits form gels.

pectins complex POLYSACCHARIDES occurring in the cell walls of certain plants particularly fruits. They are soluble in water and acid solutions, and gel with sucrose, i.e. they set to a jelly, hence their use in jam making.

pectoral girdle *or* **shoulder girdle** the skeletal structure in vertebrates that provides attachment for the anterior limbs (arms, pectoral fins, etc). It is composed of cartilage or bone, and in mammals comprises two shoulder blades (SCAPULAE) attached to the vertebral column and two collar bones (CLAVICLES) attached to the breastbone (STERNUM).

pedicel the stalk of an individual flower that attaches it to the main stem or axis.

pedology the study of soils—their composition, occurrence and formation.

peduncle in botany, the main stalk of a plant bearing several flowers. In zoology, the stalk by which certain organisms, e.g. brachiopods, anchor themselves to the substrate.

pelagic said of organisms living in the sea between the surface and middle depths. Pelagic sediments (e.g. OOZE) are deep water deposits comprising minute organisms and small quantities of fine-grained debris.

pellicle a thin surface layer of protein that covers, protects and maintains the shape of some unicellular organisms. In ciliated forms, e.g. *Paramecium*, the cilia protrude through pores in the pellicle.

pelvic girdle *or* **hip girdle** the skeletal structure in vertebrates that provides attachment for the posterior limbs (legs, pelvic fins, etc). It is composed of cartilage or bone and articulates with the vertebral column. It is made up of two halves, each comprising the ILIUM, PUBIS and ISCHIUM fused together.

pelvis the PELVIC GIRDLE. Also, the part of the abdomen in the area of the pelvic girdle and a conical chamber in which URINE collects before passing to the URETER (*see* KIDNEY).

penicillin an antibiotic, *penicillin G*, that belongs to a class of substances called penicillins. It is derived from the mould *penicillium notatum* and is active against a number of bacteria. It works through disrupting the synthesis of bacterial cell walls.

penis the reproductive organ of male mammals, and some birds and reptiles, that introduces sperm into the reproductive tract of the female to achieve internal fertilization. The sperm pass through the URETHRA, a duct that also drains the bladder. To achieve insertion into the *vagina*, the penis becomes filled with blood or haemolymph or it is erected by muscles (some mammals possess a bone—baculum—that stiffens the penis, e.g. raccoons, walruses).

pentadactyl limb the limb characteristic of tetrapod vertebrates such as mammals, reptiles, birds and amphibians. There are five digits, and the limb developed from the fins of primitive fish that were used for moving on land. There are three parts to the limb: the upper arm (or thigh); the forearm; and the hand or foot, which contains several small bones. Different species have modifications of this basic design depending upon usage, e.g. the flipper of a whale has greatly shortened upper and middle parts, and the wing of a bat has relatively long thin metacarpals and PHALANGES.

pentose a MONOSACCHARIDE with five carbon atoms in the molecule. Examples are ribose and deoxyribose (NUCLEIC ACID components). Pentoses are found in chains of polysaccharides in plants.

pepsin a proteolytic (*see* PROTEOLYSIS) enzyme that is secreted by cells lining the stomach. It is active in acid conditions and catalyses the breakdown of proteins, giving PEPTIDES and AMINO ACIDS.

peptidase enzyme(s) that hydrolyse (*see* HYDROLYSIS) peptides into constituent AMINO ACIDS (*see also* ENDO- and EXOPEPTIDASE).

peptide an organic compound made up of two or more AMINO ACIDS and linked by PEPTIDE BONDS. Peptides are collectively named by the number of amino acids present, thus a dipeptide contains two and a polypeptide contains many, perhaps up to 200 or 300.

peptide bond the chemical linkage formed when two AMINO ACIDS join together. As all amino acids have a common molecular structure, the reaction always involves the elimination of a water molecule as the amino group (NH_2) of one amino acid molecule joins to the carboxyl group (COOH) of another molecule.

perennation the means whereby biennial and perennial plants survive from year to year by vegetative means. After the aerial parts of the plant die down in biennials and herbaceous perennials, underground roots enable the plant to survive. These may be TUBERS, bulbs, CORMS, RHIZOMES, etc. The woody perennials survive winter through leaf loss, as in deciduous shrubs and trees.

perennial a plant that grows in successive years whether through addition to previous years' growth as in trees, or the cyclical dying down and regrowth of shoots each year, as in herbaceous varieties (*see* PERENNATION).

perianth the part of a flower, comprising the sepals and petals, that is outside and encloses the stamens and carpels. It has different forms in the MONOCOTYLEDONS and dicotyledons, and some varieties have been bred to produce a double flower, but such forms are often sterile.

pericardial cavity in vertebrates, the cavity containing the heart and surrounded by the double walled membrane, the pericardium.

pericarp *see* **fruit**.

periodic table an ordered arrangement of the elements by ATOMIC NUMBER. The elements are arranged by periods (horizontally, *see* APPENDIX 1), which correspond to the filling of successive shells, and by groups (vertically), which reflect the number of VALENCY ELECTRONS, i.e. the number in the outer shell.

peripheral nervous system those parts of the nervous system excluding the central nervous system, i.e. the brain and spinal cord. It consists of the two types of cells, sensory or AFFERENT and motor or EFFERENT. In humans this system includes twelve pairs of CRANIAL NERVES and thirty-one pairs of spinal nerves. The motor nervous system then consists of the somatic nervous system carrying impulses to skeletal muscles and the AUTONOMIC NERVOUS SYSTEM, which is further subdivided into the SYMPATHETIC and PARASYMPATHETIC NERVOUS SYSTEMS.

Perissodactyla a mammalian order of herbivorous animals that have hoofed feet and an odd number of toes. The teeth are specialized for grinding, and digestion of cellulose occurs in the CAECUM and large intestine. Included are the horse, rhinoceros, zebra, and tapir. This was a distinct line 60 million years ago in the Eocene.

peristalsis the involuntary muscular contractions responsible for moving the contents of tubular organs in one direction. Peristalsis occurs in the alimentary canal of animals as the alternate waves of contraction and relaxation of smooth muscle move food and waste products along.

peristome a ring of structures, resembling teeth, around the opening of a moss CAPSULE. In dry weather, the teeth bend allowing the spores out. Also, in many invertebrates, the area around the mouth which may be ciliated and often assists in gathering food.

peritoneum the lining of the vertebrate abdominal cavity. The tissue (*see* SEROUS MEMBRANE) also covers the abdominal organs.

permafrost frozen ground that is permanently frozen except for surface melting in the summer. About one quarter of the Earth's land surface is affected, and although it may be very thick (several hundred metres) the larger depths are probably relics from the last ice age. It occurs north of the Arctic Circle, in Canada, Alaska and Siberia. Permafrost can cause considerable engineering problems, particularly when the heat generated by towns, etc, creates some thawing, leading to subsidence and slumping of previously solid ground.

Permian the last geological period of the Palaeozoic extending from 286 to 248 million years ago. In many areas continental conditions resulted in the deposition of the New Red Sandstone. Among the vertebrates there were fish, amphibians and reptiles and the latter two were the dominant land animals. During the Permian the trilobites and some corals became extinct, and GYMNOSPERMS replaced the pteridophytes (ferns) as the dominant plants.

pests a species that conflicts with or interrupts human lifestyle or existence. A species can become a pest in several ways: a species may move in to new areas; there may be an increase in numbers and thus nuisance value; a change in human activities bringing into conflict humans and the pest species where previously there was little or no contact.

Pest control can take several forms including the use of specific pesticides; biological control utilizing another organism; cultural control where some practice (perhaps agricultural) is altered to affect the pest; breeding, where resistant strains are produced, and "sterile mating control," which involves releasing sterile pest individuals into the breeding population to reduce numbers.

petal part of the corolla (i.e. the ring of petals and the inner ring of the PERIANTH) of a flower. Varieties that are pollinated by insects have brightly coloured and/or scented petals while wind-pollinated varieties have reduced petals or none at all.

petiole the stalk of a leaf.

petroleum *or* **crude oil** a mixture of naturally occurring HYDROCARBONS formed by the decay of organic matter under pressure and elevated temperatures. Oil thus formed migrates from its source to a permeable reservoir rock, which is capped or sealed by an impermeable cover. The composition of the petroleum varies with the source and is separated initially by fractional distillation into its major components (gas, liquids, wax, and residues such as bitumen). The liquids include petrol, paraffin oil, and other hydrocarbon liquids. CRACKING is used to break down some substances to create smaller molecules that can be used more readily. In addition to the production of various fuels, petroleum is the basis of the vast petrochemicals industry. *See also* FOSSIL FUELS.

PG *abbreviation for* PROSTAGLANDIN.

pH the measure of concentration of hydrogen IONS (H^+) in an aqueous solution. The pH is the negative logarithm (base 10) of H^+ ion concentration, calculated using the following formula:

$$pH = \log_{10} (1/(H^+))$$

The scale of pH ranges from 1.0 (highly acidic), with

decreasing acidity until pH 7.0 (NEUTRAL) and then increasing alkalinity to 14 (highly alkaline). As the pH measurement is logarithmic, one unit of pH change is equivalent to a tenfold change in the concentration of H⁺ ions.

phage *see* **bacteriophage**.

phagocyte a cell that can surround and break down cell debris, foreign particles and micro-organisms that produce infection or disease (*see* PHAGOCYTOSIS and INFLAMMATION). For most animals, phagocytes are an important part of their natural defence mechanism. There are many phagocytic PROTOZOANS.

phagocytosis the process by which cells bind and ingest large particles from the surrounding environment. In phagocytosis, the target particle binds to the cell's surface and is then completely engulfed by a bud formed by the plasma membrane of the cell. This process is used by simple unicellular organisms to ingest food particles and by certain LEUCO-CYTES to engulf and destroy bacteria and old, broken cells.

phalanges in vertebrates, the bones comprising the digits (fingers and toes). They articulate with the metacarpals or metatarsals (for hand and foot respectively) and each finger has one to five phalanges. The PENTADACTYL LIMB has two for the first digit and three for the others.

pharynx the extent of the vertebrate gut between buccal cavity (mouth) and the OESOPHAGUS and TRA-CHEA. It is the route for food and respiratory gases and the presence of food in the pharynx stimulates swallowing. In tetrapods the glottis opens into the pharynx from the trachea, whereas in fish and aquatic amphibia the pharynx bears GILL SLITS.

phase a chemical phase is a part of a system which is chemically and physically uniform and which is separated from other such parts by boundary surfaces, i.e. a distinct interface. Thus in a system with ice, water and water vapour, there are three phases. However, all gases show one phase since they all mix with each other.

phase rule a relationship devised by Gibbs in the 19th century relating number of phases (P), to degrees of freedom (F) and the number of components (C). Thus in a chemical system

$$P + F = C + 2$$

phenol (C_6H_5OH—carbolic acid) as a solution in water, it is corrosive and poisonous. It is used as a disinfectant (with the typical "carbolic" smell) and in the manufacture of dyes and plastics.

phenotype the detectable characteristics of an organism that are determined by the interaction between its GENOTYPE and the environment in which the organism develops. Organisms with identical genotypes may have different phenotypes, because of development in environments that differ in, for example, the availability of important nutrients or specific stimuli. It is unlikely, however, that organisms that have identical detectable phenotypes will have different genotypes unless they are HETEROZYGOTES. The expression of the dominant gene masks the presence of a recessive gene, as only the expressed gene affects the organism's phenotype.

phenyl ketonuria a human genetic disorder resulting in the deficiency of the ENZYME that converts dietary phenylalanine (an AMINO ACID) to tyrosine. Children can be severely retarded (mentally), and there is an accumulation of phenylalanine and the ketone phenylpyruvate in body fluids. The disorder occurs in individuals who are HOMOZYGOUS for the defective ALLELE.

pheromone a molecule that functions as a chemical communication signal between individuals of the same species. Pheromones are used extensively throughout the animal kingdom and have a wide range of functions. They can act as sexual attractants (very common in insects) and can help establish territories, as demonstrated by the frequent urination by dogs. Although pheromones are much rarer in plants than animals, one of the most economically and environmentally important pheromones is produced by a plant called the "Scary Hairy Wild Potato" (*Solanum berthaultii*). The leaves of this plant produce a pheromone that is identical to the warning signal produced by aphids. Breeding this aphid-repellent character into cultivated crops will reduce the financial loss from crop damage and will reduce pollution as fewer insecticides will be needed.

phloem tissue in plants that has the major task of transporting food materials from the points of production, the leaves, to areas where they are needed, e.g. growing points. The phloem is made up of sieve tubes, which are hollow and lie parallel to the length of the plant. The tubes are formed from sieve cells, end to end, with end cell walls broken down to permit movement of the metabolites.

Phoronida a phylum of marine, tube-dwelling worms that are unsegmented and possess a COELOM and a lophophore (a ciliated organ that is for feeding). Some live in chitinous tubes, buried in sand. There are just two genera and fifteen species.

phosphagen a high-energy phosphate compound in animal tissues that acts as an energy store in the form of high-energy phosphate bonds. Vertebrate muscles and nerves contain creatine phosphate and invertebrates contain arginine phosphate. Creatine (a compound derived from the AMINO ACIDs, arginine, glycine and methionine) and arginine itself become phosphorylated when ATP concentration is high. Then, during tissue activity (as in muscle contraction) phosphagens lose their phosphate groups and generate ATP from ADP.

phospholipids compounds containing two FATTY ACIDs and a phosphate group based on GLYCEROL (or an alcohol sphingosine). The tails of phospholipids consist of hydrocarbons and are hydrophobic while the phosphate group makes the "head" hydrophilic and able to mix with water. Because of this, phospholipids readily form membrane-like structures and as bilayers (double layers) they form a major constituent of cell membranes. The hydrophilic top is on the outside in contact with water or an aqueous solution and the tails point into the membrane forming, overall, a boundary between the cell and the "outside."

The glycerophospholipids include LECITHIN, and the sphingophospholipids are found in animal and plant membranes.

phosphorescence LUMINESCENCE that continues after the initial cause of excitation. The substance usually emits light of a particular wavelength after absorbing electromagnetic radiation of a shorter wavelength.

phosphorylase an ENZYME that enables the transfer of a phosphate group onto an organic compound.

313

phosphorylation

The compound is then said to be phosphorylated (*see also* PHOSPHORYLATION).

phosphorylation the process of transferring a phosphate group to an organic compound by means of a PHOSPHORYLASE. It is a very important metabolic energy-transfer system. *Oxidative phosphorylation* is a reaction occurring in the inner membrane of the MITOCHONDRION. It occurs at the final stages of AEROBIC respiration and produces ATP from ADP (and phosphate) in concert with electron transport in the ELECTRON TRANSPORT CHAIN. About 90 per cent of the ATP generated by respiration is generated by oxidative phosphorylation. A small amount of ATP is produced in reactions of glycolysin and the KREBS CYCLE through *substrate level phosphorylation* when a phosphate is moved from a substrate to ADP.

photic zone the uppermost layer of a lake or sea where there is adequate light to allow PHOTOSYNTHESIS to proceed. The limit will vary depending on the quality of the water and the material held in suspension, but can be as much as 200 metres.

photobiology the study of the effect of light on living organisms.

photolysis the decomposition or reaction of a substance because of the absorption of light or ultraviolet radiation. Flash photolysis is a technique for studying very fast reactions, involving ATOMS or RADICALS in the gas phase. The reactants are subjected to an intense, but brief, flash of light, which causes dissociation. Subsequent flashes are used immediately afterwards to identify intermediates produced in the reaction (by studying the absorption spectra).

photon a QUANTUM of energy that is an intrinsic

component of all ELECTROMAGNETIC WAVES. Photons are used to explain the quantum theory of light, where the properties of light are explained in terms of particles (photons), as opposed to the wave theory of light, where its properties are explained by the propagation of a wave and how it disturbs a medium. The energy of a photon is proportional to the FREQUENCY of the light beam.

photoreceptor a cell, or cells, that reacts to light by means of a pigment that undergoes chemical change, stimulating a nerve (*see also* EYE, CONE and ROD CELL).

photosynthesis the process by which plants make carbohydrates, using water, carbon dioxide (CO_2) and light energy, while releasing oxygen. Photosynthesis occurs in two stages, known as the CALVIN CYCLE and the LIGHT REACTIONS of photosynthesis. For photosynthesis to occur, an organism must contain light-trapping pigments, which capture light energy in the form of PHOTONS and use the photons to initiate a series of energy-transfer reactions. Some blue-green algae (cyanobacteria) and the CHLOROPLASTS of all plants contain the essential light-trapping pigment called CHLOROPHYLL that makes them capable of photosynthesis. Photosynthesis is an essential process for regulating the atmosphere as it increases the oxygen concentration while reducing the CO_2 concentration.

phototaxis the movement or reaction of an organism in response to light (*see also* CHEMOTAXIS).

phototropism a growth movement exhibited by parts of plants in response to the stimulus of light. Plant shoots display positive phototropism as they grow towards the light source, but the roots tend to display negative phototropism as they grow away from the light source (*see* GEOTROPISM). Phototropism is caused by the unequal distribution of auxin (a plant growth hormone) as this substance has a higher concentration in the darker side of the plant and thus increases growth on this side by inducing cell elongation.

phylloquinone *see* **vitamin K**.

phyllotaxy the way in which leaves are arranged on a plant stem, whether singly, in pairs or rings, at each node. Paired leaves, on either side of the stem, are usually at 90° to pairs above and below. There is a numerical system (the Fibonacci series) which characterizes each form. The simplest is one leaf above which is another at 180°, such that one "revolution" of the stem encompasses two leaves, a phyllotaxy of 1/2. There are various forms: 1/3, 3/8, etc, depending upon the arrangement seen when looking vertically down the stem. The main result of phyllotaxy is that higher leaves shade those below as little as possible.

phylogenetics biological classification of organisms showing their evolutionary history, i.e. it recreates PHYLOGENY.

phylogeny the evolutionary history of an organism or a group that is closely related.

phylum (*plural* **phyla**) a part of the taxonomic classification of the animal kingdom. A phylum includes one or more classes that are closely related. Examples are Protozoa, Arthropoda, Chordata. In plant classification, the term "division" is used.

physical chemistry the study of the link between physical properties and chemical composition and the physical changes caused by chemical reactions.

physiology the study of all functions of animals and plants including the processes of respiration, reproduction, nutrition, etc, involving cells, tissues and organs.

phytoalexins compounds in plants that are instrumental in disease resistance. In the main phenolic (*see* PHENOL) or TERPENOID (*see also* TERPENES), they are produced or increased in concentration to restrict the growth of, or destroy, fungi.

phytochemistry the study of the chemical make-up of plants.

phytochrome a plant pigment, based on protein, that is present in small amounts, and occurs in two forms that can interchange. When a plant absorbs red light or normal daylight phytochrome red forms, while in far-red light or darkness, phytochrome far-red is formed. The red form is the active version, controlling flowering and seed germination. When the red form absorbs light it converts to the far-red form and when in the dark the change is reversed to create the red form again.

phytoplankton the plant part of PLANKTON that is mainly algae (DIATOMS and dinoflagellates). These form the basis of food for many aquatic organisms and occur in enormous numbers.

pi (π) bond the COVALENT BOND formed when two atoms join to form a diatomic molecule. Pi bonds are discussed in terms of molecular ORBITALS, in which the shared electrons orbit the whole molecule rather than an atom. Pi bonds hold the molecule together by forming two regions of electron density above and below an axis between the bonded nuclei of the two atoms.

pineal gland in mammals, a small mass of nerve

tissue near to the centre of the brain. It secretes the HORMONE melatonin which seems to be involved in CIRCADIAN RHYTHMS. However, its precise role is unclear.

pinna the "ear" of mammals, i.e. the visible part of the outer ear, which is made of cartilage and serves to direct sound waves into the ear.

pinocytosis the surrounding and engulfing of a drop of fluid by a cell forming a vesicle.

Pisces fishes—cold-blooded aquatic vertebrates with GILLS and FINS. The term is no longer used and modern fish are divided into two classes, the CHONDRICHTHYES and Osteichthyes (*see* BONY FISHES).

pistil the female part of a flower. It may be a single or compound structure and typically contains the ovary, style and stigma.

pituitary gland *or* **hypophysis** a small gland at the base of the HYPOTHALAMUS that has two lobes, the anterior and posterior or *adenohypophysis* and *neurohypophysis* respectively. The pituitary secretes HORMONES that control many functions and it is itself controlled by hormonal secretions from the hypothalamus. The posterior lobe releases peptide hormones that are produced by the hypothalamus, namely oxytocin (which stimulates uterine contraction, *see* PARTURITION) and VASOPRESSIN.

The anterior lobe secretes a number of hormones including GROWTH HORMONE, GONADOTROPHIN, PROLACTIN, ACTH and THYROID stimulating hormones (*see* ENDOCRINE SYSTEM).

placenta the mammalian organ, found also in other VIVIPAROUS animals, that attaches the embryo to the uterus. It is a temporary feature made up of maternal and embryonic tissues and allows oxygen and nutrients to pass from the mother's blood to the embryo's blood although there is no direct contact of blood supplies. In addition to oxygen, the embryo receives salt, glucose, amino acids and small peptides, some antibodies, fats and vitamins. Waste molecules from the embryo are removed, mainly carbon dioxide and urea, by diffusion into the maternal circulation. The placenta also stores glycogen for conversion to glucose if required by the embryo. The placenta is expelled after birth.

plagiotropism when a plant growth response is at an angle to the direction of the applied stimulus.

plankton very small organisms, of plant and animal origins, that drift in water. The plants (or phytoplankton) are mainly diatoms (unicellular ALGAE) that PHOTOSYNTHESIZE and form the basis of the food chains. The animals (or zooplankton) feed on the diatoms and include small crustaceans and the larval stages of larger organisms.

plasma in biology, the same as BLOOD PLASMA. In physics, essentially a high temperature gas of charged particles rather than neutral atoms or molecules.

plasmalemma, plasma membrane *see* **cell membrane**.

plasmid a small molecule of DNA, separate from the CHROMOSOMES of cells, that can replicate independently (*see* REPLICATION). It is an extra chromosomal DNA that contains a few genes that sometimes confer a particular property on the host cell, e.g. drug resistance. Bacterial forms are used in gene cloning.

plasmolysis *see* **osmosis**.

plastid small, often numerous, organelles in plant

cells containing pigment and/or food material. When mature they are identified by pigments. CHROMO-PLASTS contain pigments and leucoplasts do not, although the two may interchange.

platelet *see* **blood**.

Platyhelminthes an invertebrate phylum, containing the flatworms, which have a flat, non-segmented body. There are about 20,000 species in the phylum, occupying marine, freshwater and damp terrestrial habitats. There are parasitic forms (tapeworm and the fluke). The phylum is divided into four classes: TURBELLARIA (free-living worms); CESTODA (tapeworms); TREMATODA, and Monogenea (flukes).

pleiomorphism when an individual exists in different forms throughout its life cycle, e.g. caterpillar, pupa and butterfly.

Pleistocene an epoch of the QUATERNARY extending from two million years to 10,000 years ago. Often called the ice age, during which glaciers extended southwards in advances punctuated by retreats. There was rapid evolution of the hominids, and the elephant, cattle and modern horse all appeared.

pleura (*plural* **pleurae**) a double SEROUS MEMBRANE lining the thoracic cavity and covering the lungs.

plexus a network of branching nerves or blood vessels, e.g. the choroid plexus, a membrane in the brain rich in blood vessels, and the brachial plexus (spinal nerves in the forelimbs of vertebrates).

Pliocene the last epoch of the TERTIARY, from 5.1 to two million years ago. The early forerunners of man, the australopithecines, appeared, and mammals similar to modern forms existed.

podsol a soil with minerals leached from its surface layers into lower layers. Podsolization is an advanced stage of leaching that involves the removal of iron and aluminium compounds, humus and clay minerals from the topmost horizons and their redeposition lower down.

poikilothermy *or* **cold-bloodedness** a feature of all animals except mammals and birds. The body temperature of poikilotherms varies with the surrounding temperature and although control is minimal, animals can seek sun or shade as required. In addition, blood flow to peripheral tissues can be changed (*see also* HOMOIOTHERMY).

point of inflection a point where a plane curve changes from the concave to the convex,

polar covalent bond the joining of two ATOMS due to the strong but unequal sharing of their ELECTRONS, which gives the bond, and thus the molecule formed, partial charges. Polar covalent bonds are formed between atoms that differ in their ability to attract electrons, with one atom having a greater ELECTRONEGATIVITY than another. For example, a hydrogen chloride molecule (HCl) has a polar covalent bond, as the chlorine atom is more electronegative than the hydrogen atom. The net result is that the chlorine end of the bond has a denser electron cloud because of its greater attraction for electrons. This causes the chlorine end of the molecule to have a partial negative charge, and as the whole HCl molecule is neutral, the hydrogen end has a corresponding partial positive charge.

polarimetry the measurement of OPTICAL ACTIVITY in, for example, sugar solutions. Also the measurement of the extent to which light reflected from planetary surfaces is polarized.

pollen the male gametes of seed plants found in sacs

(pollen sacs) in the ANTHER. The pollen grain is the male GAMETOPHYTE generation. The form of a grain varies with the mode of pollination of the respective plant. Wind-pollinated plants have light, smooth grains while insect-pollinated plants have grains that are rough or spiny.

pollen analysis (*see also* PALYNOLOGY) a useful tool in reassembling the history of the flora of an area. Because the outer layer of pollen grains is resistant, particularly if deposited under the anaerobic conditions of rapid sedimentation, or in peat or stagnant water, they are widely distributed and preserved. Such analysis contributes to studies of climate change and sediment dating.

pollination the transfer of pollen from the ANTHER to a STIGMA. The agent may be wind, insect, bird, water, etc. Self-pollination is where anther and stigma are on the same flower, cross-pollination a different flower of the same species.

Polychaeta a class of worms within the phylum ANNELIDA. Each segment of the body has a pair of lobes (*parapodia*) with many bristles (CHAETAe). In many cases the parapodia function as gills and in locomotion. Most polychaetes are marine and many inhabit tubes constructed from sand and shell pieces. Included in the latter group are fan worms (*Sabella*). Lugworms (*Arenicola*) and ragworms (*Nereis*) burrow in mud or sand. There is usually a distinct head, often with jaws and eyes. The sexes are separate, and fertilization is internal.

polymerase an ENZYME that catalytically joins together monomers to form a polymer. RNA polymerase catalyses RNA synthesis using existing RNA or DNA as a template.

polymorphism when three or more clearly different forms of a plant or animal species exist, e.g. the worker, drone and queen honey bees.

polyp the fixed stage in the life cycle of the CNIDARIA (coelenterates). The polyp is cylindrical and the free end bears a mouth surrounded by tentacles. Examples are *Hydra* and sea anemones. Polyps often reproduce asexually by BUDDING although some are sexual, e.g. the sea anemones. Some polyps form MEDUSAe and the latter can reproduce sexually to produce more polyps.

polypeptide a single, linear MOLECULE that is formed from many AMINO ACIDs joined by PEPTIDE BONDS. Polypeptides differ greatly in the number of amino acids they contain (usually from 30 to 1000). Although there are only 20 different amino acids, there are a huge number of possible arrangements in a polypeptide or PROTEIN, as the amino acids can be in any order. Most proteins consist of more than one polypeptide rather than a single polypeptide chain.

polyphyletic when a group of organisms has evolved from more than one, different, ancestor. This is not a natural classification but is manifested in the group that includes the insectivores.

pome a false fruit (*see* FRUIT), such as the pear or apple where the fruit flesh grows from the receptacle of the FLOWER and not the ovary. In this case, the core is the ovary.

pons *or* **pons Varolii** nerve fibres that link the MEDULLA OBLONGATA to the MIDBRAIN. The pons, named after the Italian anatomist C. Varoli who discovered it, passes impulses between different parts of the brain.

population a group of individuals that belongs to the same species and that is subject to VARIATION because of the effect of factors such as birth and death rates, population density, immigration and emigration. Populations may be isolated, rarely exchanging genetic material, or one dense population may merge into another.

population dynamics the study of the changes that occur in population numbers, whether plant or animal, and the factors that control or influence these changes. Distinction is made between factors that are dependent or independent of population density. A natural occurrence, e.g. a flood, is an independent factor while food supply is dependent.

population genetics the study of (inherited) VARIATION among a group of organisms within the same species. The possibility of change depends upon the total ALLELES available to the population. The total of all the genes in a population at one time is called the *gene pool* (*see also* HARDY-WEINBERG RATIO).

Porifera a phylum of invertebrates—the sponges. Only about 100 of the 9000 or so species live in freshwater, the remainder being marine. Sponges live attached to rocks or other substrates and essentially resemble a hollow sac-like shape that is perforated with holes. The body has an internal skeleton made up of spicules of calcium carbonate or silica (or in some cases protein). Cells with FLAGELLAE (*choanocytes*) make water flow through the pores into the central cavity (*spongocoel*) and out through the opening at the top of the sac. Sponges are filter-feeders, and the choanocytes take particles of food out of the water, which are then ingested by PHAGO-CYTOSIS. Sponges are mainly hermaphrodite, producing eggs and sperm and then flagellated larvae. Upon settling on a substrate the larva metamorphoses, turning inside out. Sponges are very simple organisms with no muscle or nerves, and they are able to regenerate lost parts.

porphyrins naturally occurring pigments that include CHLOROPHYLL and the hæm part of HAEMOGLOBIN.

portal vein a general term for a vein that connects two networks of capillaries, delivering blood from one to the other, e.g. HEPATIC PORTAL SYSTEM.

posterior the part of an animal to the rear, which in humans is synonymous with the DORSAL surface.

potassium (K) a soft alkali metal that is highly reactive. Combined with other elements it occurs widely, in silicate rocks, in blood and milk and also in plants. Potassium is an essential element for living organisms and the cation K⁺ is the most abundant in plant tissues. In animals potassium, in concert with sodium, is responsible for changes in electrical potential during the transmission of nerve impulses. In the human diet, potassium is obtained from milk, meat and fruits and in addition to nerve function it is instrumental in maintaining body water and acid-base balance. Deficiency may lead to muscle weakness.

potential evapotranspiration the maximum quantity of water vapour that, in theory, can be released into the atmosphere by both evaporation and transpiration from an area of green vegetation that is fully supplied with water.

Precambrian an informal name attributed to geological time from the formation of the Earth (about 4600 million years ago) to the CAMBRIAN (590 million years ago). Because of the age, and therefore extensive alteration, of these rocks, few fossils are preserved, although STROMATOLITES and RADIOLARIA are found.

premolar a broad tooth in mammalian DENTITION, behind the CANINE tooth and in front of the MOLARS. Premolars are used for grinding and chewing food.

Primates a mammalian order that includes monkeys, lemurs, apes and humans. Evidence suggests that primates evolved from tree-dwelling insectivores late in the CRETACEOUS. Characteristic features are manual dexterity (permitted by thumbs and big toes that can touch and grasp), binocular vision, and good eye to hand co-ordination. Parental care is protracted. The brain, and in particular the CEREBRUM, is large and explains the intelligence exhibited by this order. There are two suborders, the Prosimii (lemurs) and the Anthropoidea (monkeys, apes and humans). It seems that the Anthropoids developed from prosimian ancestry about 40 million years ago.

primordium (*plural* **primordia**) a cell or group of cells in a plant that eventually develops into a specific structure.

Proboscidea the order of placental mammals that consists of the elephants, which evolved during the Eocene and was larger in numbers than today, containing as it did the now extinct mammoths. Elephants are herbivorous and have the characteristic trunk (PROBOSCIS) and tusks, which are modified incisors. There are just two modern species, the African and Indian.

proboscis the elongated mouthparts of some flies and most butterflies and moths. Also the trunk of an elephant.

procambium plant tissue comprising long cells, elongated parallel to the axis of the plant, which gives rise to the VASCULAR TISSUE.

procaryote any organism that lacks a true-membrane NUCLEUS and is either a bacterium or a blue-green algae (cyanobacteria). Procaryotes have a single CHROMOSOME and do not undergo MEIOSIS or MITOSIS as they lack the MICROTUBULES to form the spindle. Procaryotes replicate by a form of asexual reproduction, called binary fission, in which the two sister chromosomes are attached to separate regions on the cell membrane, which starts to fold to form a cleavage. The cell eventually forms two daughter cells after the CYTOPLASM has been completely split by the fusion of the enfolding cell membrane.

profundal the area of an inland lake that is the deep water zone, where there is low oxygen concentration, light levels and temperatures compared with the upper layers (e.g. LITTORAL).

progesterone a STEROID hormone that is vital in pregnancy. The HORMONE is produced by the CORPUS LUTEUM of the OVARY, when the uterine lining is prepared for the implantation of an egg cell. Under the control of PROLACTIN and LUTEINIZING HORMONE the corpus luteum secretes progesterone until, some months into the pregnancy, the PLACENTA takes over the function. The hormone maintains the uterus and ensures the ovary produces no further eggs. Small amounts of progesterone are produced by the male testes.

prolactin *or* **luteotropic hormone** (LTH) or **lactogenic hormone** in vertebrates, a HORMONE secreted by the anterior PITUITARY GLAND. In addition to prompting the corpus luteum to secrete PROGESTER-ONE, it also stimulates lactation, i.e. production of milk by the MAMMARY GLANDS. Prolactin secretion is enhanced by suckling. In birds, it prompts the crop glands to produce crop milk (*see* CROP).

propanone *see* **acetone**.

prophase the first stage of MEIOSIS or MITOSIS in cells of EUCARYOTES. During prophase, the CHROMOSOMES condense and can thus be studied using a microscope.

proprioceptor a RECEPTOR that is involved in the detection of movement and position. They occur in muscles where they are essential to muscular activity and posture and also in the INNER EAR as part of the mechanism conferring balance.

prosencephalon *see* **forebrain**.

prostaglandin (PG) a group of organic compounds derived from ESSENTIAL FATTY ACIDS. Prostaglandins are found in most body tissues, being released by most cells into the interstitial fluid where they act as local regulators. Sixteen have been identified, and two act antagonistically on blood vessels serving the lung; PGE dilating and PGF constricting the vessels. Certain PGs are released by the placenta to cause uterine contraction during labour. Still others are active in the body's defence mechanisms.

prostate gland a gland in the reproductive system of male mammals. The gland is below the bladder and opens into the URETHRA. Upon ejaculation, it secretes an alkaline fluid into the semen. The fluid aids sperm motility and makes up one third of the volume.

protandry in flowers, the term for STAMENS (male) maturing before the CARPELS (female). This ensures that self-fertilization cannot occur.

protease a group of ENZYMES that act as catalysts in the breaking up of PROTEINS into PEPTIDES and AMINO ACIDS. Examples are PEPSIN and trypsin.

protein a large group of organic compounds containing nitrogen and usually sulphur, with individual molecules built up of AMINO ACIDS in long POLYPEPTIDE chains and with MOLECULAR WEIGHTS ranging up to several million. The amino acids of each polypeptide confer upon it its *primary structure*, which may be coiled to form the *secondary structure* (because of HYDROGEN BONDING). The corkscrew of an α (alpha)-helix is a secondary structure. The combination of α-helix structures with pleated sheets and more random chains produces a three-dimensional configuration—the *tertiary structure*. The way polypeptide chains adopt a particular shape in a functional protein is then the *quaternary structure*.

Of the two types of protein, globular includes ENZYMES, ANTIBODIES, carrier proteins (e.g. HAEMO-GLOBIN), some HORMONES and proteins such as CASEIN and ALBUMIN. Fibrous proteins possess strength and elasticity and include those in muscle, connective tissue and also chromosomes.

When subjected to heat (over 50°C) proteins denature and lose their structure and some coagulate (as with the white of an egg).

protein synthesis the manufacture of PROTEINS from AMINO ACIDS by living cells. The necessary "instructions" are carried in chromosomal DNA, encoded in MESSENGER RNA and the sequence of amino acids is dependent upon the NUCLEOTIDE sequence in the messenger RNA.

proteolysis the splitting of proteins through the catalytic action of PROTEASES. To break down a protein completely (into its amino acids) usually requires several proteases acting one after the other.

Protista a biological kingdom containing numerous phyla with simple organisms such as *Amoeba*, the RADIOLARIA, FORAMINIFERA, the various types of ALGAE, SLIME MOULDS and others. The protists are found just about anywhere there is water forming major components of plankton, and they also occur in damp terrestrial habitats. Most are ciliated or flagellated (*see* CILIA and FLAGELLUM) and AEROBIC, but in general there is such diversity that few characteristics apply to all members of the kingdom (*see also* PROTOZOA).

protogyny in flowers, the term for the CARPELS (female) maturing before the STAMENS (male). As with PROTANDRY, this avoids self-fertilization.

proton a particle that carries a positive charge and is found in the NUCLEUS of every ATOM. As an atom is electrically neutral, the number of protons equals the number of negatively charged ELECTRONS. Although the mass of a proton (1.673×10^{-27} kg) is far greater than the mass of an electron (9.11×10^{-31} kg), their charges are equal in magnitude. The number of protons in the nucleus of an atom (ATOMIC NUMBER) is identical for any one element and is used to classify elements in the PERIODIC TABLE. For example, as every oxygen atom contains 6 protons, it has an atomic number of 6 in the periodic table, whereas every gold atom contains 79 protons and thus has an atomic number and periodic table position of 79.

protoplasm the living contents of a cell including the plasma membrane but excluding large vacuoles. For most EUCARYOTIC cells it includes the NUCLEUS and CYTOPLASM.

Prototheria a mammalian subclass that contains two groups of *monotremes*, the platypuses and echidnas (spiny anteaters). These are the only mammals living that lay eggs and today are found only in Australia and New Guinea. The monotremes resemble reptiles in their skeleton and the laying of eggs and although they are warm blooded, the body temperature is variable and lower than that of most mammals. It is thought that the monotremes originated about 150 million years ago early in the development of the mammals.

Protozoa formerly a phylum of UNICELLULAR or acellular (mainly) micro-organisms but now used more as a general term for protists (*see* PROTISTA) that ingest food. As such there are numerous phyla that fit this description, including the Rhizopoda (e.g. *Amoeba*), FORAMINIFERA, Ciliophora (e.g. Paramecium) and Zoomastigophora (e.g. *Trypanosoma*, *see* TRYPANOSOMES). They are very widely distributed in aquatic and damp terrestrial habitats. Some are saprophytic (absorb dead organic matter) while others are parasitic (e.g. *Trypansoma*, which causes sleeping sickness). Reproduction is usually asexual, by binary fission (*see* ASEXUAL REPRODUCTION).

provirus a stage in the infection of a cell by a RETROVIRUS where the single strand of RNA in the virus becomes double stranded DNA through the action of the enzyme REVERSE TRANSCRIPTASE (which also destroys the RNA). The DNA of the provirus then becomes part of the host cell.

proximal a term signifying near the point of attachment, e.g. the part of an organ nearest to its point of attachment.

pseudocoelomates animals that possess a body

cavity that, unlike a COELOM, is not lined with mesodermal tissue. Within this grouping are several
phyla including rotifers (phylum Rotifera) and the
roundworms (phylum NEMATODA). The position of
the pseudocoelomates with respect to other groups
is uncertain in evolutionary terms, and the condition may have several origins.

pseudopodium (*plural* **pseudopodia**) the temporary projection from the body of certain cells.
Pseudopodia are formed in simple, single-celled
organisms, such as *Amoeba*, as a mechanism for
locomotion and food intake. They are also formed by
white blood cells, which use PHAGOCYTOSIS to ingest
particles.

Pterophyta a division within the plant kingdom
containing the ferns. There are about 12000 species
of ferns which occurred as far back as the CARBONIF-
EROUS. Ferns are almost exlusively terrestrial and
the majority occur in tropical regions. The leaves
are usually called *fronds* and each contain numerous smaller leaflets and these comprise the SPORO-
PHYTE generation. Spores are released from SPOR-
ANGIA on the underside of certain leaves and are
spread by the wind. The spore develops into a
GAMETOPHYTE that photosynthesizes and produces
both eggs and sperm, the latter "swimming" through
moisture to reach and fertilize the egg. The resulting zygote then generates a new sporophyte.

PTH *abbreviation for* parathyroid hormone (*see*
PARATHYROID GLANDS).

ptyalin a carbohydrate-digesting ENZYME (an AMYLASE)
that is present in the saliva of some mammals and
is responsible for the commencement of STARCH digestion.

pubis (*plural* **pubes**) one of three bones, and the most anterior, that makes up each half of the PELVIC GIRDLE. The other bones are the ILIUM and ISCHIUM.

pulmonary artery one of the two arteries that carry deoxygenated blood from the HEART to the lungs, where it is oxygenated. The pulmonary arteries are the only ones that carry blood with a high concentration of carbon dioxide rather than a high concentration of oxygen. All other arteries carry oxygenated blood to the tissues, where oxygen is exchanged for carbon dioxide.

pulmonary vein one of the four veins that carry oxygenated blood from the lungs (two veins leave both the left and right lungs) to the left ATRIUM of the HEART. The pulmonary veins are unique, as they carry oxygenated blood while all other veins carry deoxygenated blood back to the heart after it has exchanged oxygen for carbon dioxide in the various tissues of the body.

pulp cavity the inner cavity of a vertebrate tooth, which is connected with the surrounding tissue. It contains the pulp, which is connective tissue embedded with nerves and blood vessels and lined with ODONTOBLASTS.

pulse the number of heart beats per minute, caused by the pressure of blood pumped from the heart through the arteries. The elastic arteries stretch from the pressure and in humans this can be felt where arteries are close to the skin.

pupa the stage in the life cycle of some insects between larva and adult. During this phase, METAMORPHOSIS occurs and other activities cease.

purine one of the two different structures that form the base components of DNA and RNA. A purine has a double ring structure that consists of both carbon and nitrogen atoms. The bases, ADENINE and GUANINE, are both purines that will form hydrogen bonds with their complementary PYRIMIDINE bases to form the double helix of the DNA molecule.

putrefaction the breakdown (decomposition) of plants and animals after death by anaerobic bacteria.

pyrimidine one of the two different structures that form the base components of DNA and RNA. A pyrimidine has a single ring, consisting of both carbon and nitrogen atoms. The bases, cytosine, THYMINE and URACIL, are all pyrimidines.

pyruvate a colourless liquid formed as a key intermediate in the metabolic process of GLYCOLYSIS and the production of ATP.

Q

quadrat a small sampling plot (usually one square metre), randomly chosen, within which organisms found are collected, measured and studied to gain insight into the ecological status of the area as a whole. This may be done in relation to a particular species or several different types of organisms.

qualitative analysis the chemical analysis of a sample to identify one or more constituents.

quantitative analysis determination of the relative amount of species making up a sample, which usually refers to elemental analysis.

quantum (*plural* **quanta**) a small, discrete quantity of radiant energy. Electromagnetic radiation is explained in terms of small particles as well as waves, as it is assumed that it can be absorbed or emitted in quanta. The energy of one quantum (E) is derived from the equation, $E = hv$, where v is the frequency of the radiation and it is Planck's constant.

quantum numbers a set of four numbers used to describe atomic structures (*see* ORBITALS). The first, n, the principal quantum number, defines the shells (stationary orbits in Bohr's theory), which are visualized as orbitals. The orbit nearest the NUCLEUS has n = 1, and contains two ELECTRONS. The second shell, n = 2, contains 8 electrons, the maximum number of electrons in each shell being limited by the formula $2n^2$; the orbital quantum number, l, defines the shape of the orbits within one shell, which are designated s, p, d, f orbits; the magnetic orbital quantum number, m, which sets the spatial position of the orbit within a strong magnetic field; and s the spin quantum number, based upon the assumption that no two electrons may be exactly alike, and thus opposite spins are invoked for pairs of electrons.

Quaternary a geological period, subdivided into the PLEISTOCENE and HOLOCENE epochs, that began about two million years ago and continues to the present day. It followed the TERTIARY and is the second period of the Cenozoic era.

quinine a colourless ALKALOID with a very bitter taste, which was used in the treatment of malaria.

R

race a grouping used in the classification of organisms to denote individuals within a species that are different in one or more ways from other members of the same species. They may be different because they occupy another geographical area, perhaps showing a variation in colour, or there may be behavioural, physiological or chromosomal distinctions between races. It is sometimes used in the same way as the term sub-species.

racemose inflorescence *or* **indefinite inflorescence** a type of flowering shoot in which new flowers are continuously formed by the growing tips. As a result, the oldest flowers are at the base and the youngest are at the top as in a *spike* (e.g.plantain) and a *raceme* (e.g. lupin). Or, in a flattened inflorescence, the oldest flowers can be found on the outside and the youngest ones on the inside. Various types of this occur; a *capitulum* (e.g. daisy), *umbel* (e.g. hogweed) and *corymb* (e.g. candytuft) A catkin and spadix are also kinds of racemose inflorescence.

radial symmetry the structural arrangement of an organism (or organ) so that a plane bisecting the structure in any direction gives equal and opposite halves—mirror images. Obvious examples include the coelenterates (jellyfish) and echinoderms (starfish and sea urchins) and also plant stems and roots. When applying this concept to flowers the term actinomorphy is used.

radiation the emission of energy from a source, applied to electromagnetic waves (radio, light, X-rays, infrared, etc), particles (α, β, protons, etc), and sound.

radical a group of atoms (within a compound), usually unable to exist independently, which is unchanged in reactions affecting the rest of the molecule. (Now often referred to as a group.)

radioactivity the emission of α or β particles and/or γ rays by unstable elements, while undergoing spontaneous disintegration.

radiobiology the study of biological systems as influenced and affected by radioactive materials, and the use of carefully controlled amounts of radioactive substances (tracers) to study metabolic processes.

radiocarbon dating a method of dating organic material, although it is only applicable to the last 6000–8000 years. There is a small proportion of radioactive ^{14}C in the atmosphere, which is taken up naturally by plants and animals. When an organism dies, the uptake ceases and the ^{14}C decays with a HALF-LIFE of 5730 years. Comparison of residual radioactivity with modern standards enables an age to be calculated for a sample.

radiography the process of producing an image of an object on photographic film (or on a fluorescent screen) using X-rays (or a similar short wavelength electromagnetic wave, e.g. gamma rays). The photograph thus produced is termed a radiograph and the process is used widely in diagnostic medicine.

Radiolaria a marine, planktonic order of protozoan animals belonging to the class Actinopodea. They have a spherical body shape with an outer CORTEX and inner MEDULLA separated by a membranous capsule composed of CHITIN. There are numerous pores through the capsule so that the cytoplasm of each part remains in contact. The outer cortex is jelly-like and contains numerous vacuoles and the inner medulla has one or more nuclei. The Radiolaria characteristically possess a beautiful, intricate skeleton often composed of silica. This is perforated to make a large variety of patterns and may be in the form of a lattice of spheres, one inside the other. Often there are many spines or hook-like projections radiating outwards. Spines or spicules may also be present, dispersed in the cytoplasm. Reproduction is usually by division into two, sometimes with both daughter cells having half the skeleton and regenerating the other half. In other species, one daughter cell is freed from the skeleton of the parent and grows a new skeleton of its own. Special PSEUDOPODIA (cytoplasmic projections) are characteristically present. These are known as axopodia and are used for feeding and buoyancy. When Radiolarians die, their skeletons, which are highly resistant, settle on the ocean floor to accumulate as RADIOLARIAN OOZE. When compressed to rock it forms flint and chert. Radiolarian fossils are very important and among the few found in PRECAMBRIAN rocks.

radiolarian ooze a deep sea ooze, which contains a significant proportion of radiolarian tests (protective "shells" made of silica). The ooze may be hundreds of metres thick and it covers about 5 per cent of the ocean floor.

radiolysis the chemical decomposition of a substance subjected to ionizing radiation.

radionuclide any ISOTOPE of an element that undergoes natural radioactive decay.

radius (*plural* **radii**) present in the forelimb of tetrapod vertebrates, this is the smaller of two long bones that articulate at the elbow with the humerus and at the wrist with the carpals (*see* PENTADACTYL LIMB and ULNA).

radula this is a horny strip present in such animals as CEPHALOPODS and MOLLUSCS, which acts as a tongue and bears teeth on its surface for rasping food. It is continually renewed and may be modified for boring in some species. The pattern of teeth is a characteristic used in identification.

Ratitae the "walking birds" of the super order Palaeognathae of the class Aves, characterized by a lack of flight. They have a flat breastbone without a KEEL and heavier bones than those of other birds, with long legs and small wings. They have curly feathers, often much prized by man, and there are a number of orders within this group that represent several independent lines of evolution. Examples are ostriches, emus, kiwis, elephant birds and rheas.

reaction time also known as the latent period. This is the length of time between a stimulus being received by a sensory RECEPTOR and an appropriate response being elicited in the EFFECTOR organ. It represents the time taken for the stimulus to travel and cross one or more SYNAPSES. If few synapses are involved, as in a REFLEX response, the reaction time is very short.

receptor an excitable cell that is specialized to be sensitive to a particular stimulus within an animal's internal or external environment. When it is stimulated it sends electrical impulses via a sensory nerve involving various parts of the nervous system. Receptors may be grouped together within particular specialized sense organs, e.g. the eyes and ears, and they may be located internally or externally. They are often distinguished according to the sense that they detect, e.g. chemoreceptor (chemicals), photoreceptor (light), mechanoreceptor (touch) and proprioceptor (pressure, movement, stretching within the body).

recessive allele a gene form that is not expressed and will therefore not affect the PHENOTYPE of the organism unless the organism is HOMOZYGOUS for the recessive allele. Although an organism that is HETEROZYGOUS for a recessive allele will possess this allele, the dominant form of the gene will be expressed, thus masking the presence of the recessive form.

reciprocal cross used in genetics to confirm the results obtained from an earlier cross (mating) but this time reversing the source of male and female GAMETES. It is used to study the inheritance of certain characters and if in the first cross a male of type A was crossed with a female of type B, the reciprocal cross would be between a male of type B and a female of type A. When consistently different results are obtained in the offspring of such crosses, it may indicate that the character is sex linked in some way.

recombinant DNA a new DNA sequence formed by the insertion of a foreign DNA fragment into another DNA molecule. Recombinant DNA is used extensively throughout GENETIC ENGINEERING, when bacteria are frequently used as hosts for the expression of recombinant DNA molecules and the subsequent coding for the desired protein. It is particularly useful for producing a significant quantity of a human PROTEIN, such as INSULIN.

recombination *see* **genetic recombination**.

rectum the end part of the ALIMENTARY CANAL before the anus and after the colon where faeces are stored prior to elimination. In some aquatic insect larvae, tracheal gills are located in the rectum and in dragonfly larvae, jet propulsion is brought about by the rapid expulsion of water from this region. The rectum may reabsorb salts, water and amino acids in some insect species.

red blood cell *see* **erythrocyte**.

redox reaction a chemical reaction in which both REDUCTION and OXIDATION are involved. If the overall redox potential for such a reaction has a positive value, then it is a spontaneous and feasible reaction.

reducing agent any substance that will lose ELECTRONS during a chemical reaction. Reducing agents will readily cause the REDUCTION of other atoms, molecules or compounds, depending on the strength of the reducing agent and the reactivity of the other reactant. The strongest reducing agents are active alkali metals such as lithium (Li), potassium (K), barium (Ba), and calcium (Ca).

reduction any chemical reaction that is characterized by the loss of oxygen or the gain of ELECTRONS from one of the reactants. A molecule is also said to be reduced if it has gained a hydrogen atom. There is always simultaneous OXIDATION if reduction has occurred in any reaction.

reflex an innate, automatic and rapid response to a particular stimulus, which in its simplest form involves only a RECEPTOR and sensory NEURON synapsing with a motor neuron within the CENTRAL NERVOUS SYSTEM. This then stimulates the EFFECTOR organ (e.g. a muscle). Frequently in vertebrates there is other neuron activity·involved within the brain, enabling learning to take place as in CONDITIONED REFLEXes. If only the spinal cord is involved and not the brain, it is known as a spinal reflex. Examples of reflexes in humans are blinking, coughing and VENTILATION.

refractory period the time taken for nerve and muscle membranes to recover their resting potential after the passage of an impulse. Another ACTION POTENTIAL cannot pass along due to increased permeability of the membrane to potassium ions.

regeneration the repair or regrowth of bodily parts of an organism that have been damaged and subsequently lost. Regeneration is rare in higher, complex animals but is quite common in lower, simpler animals, in which the extent of regeneration can range from limb regeneration in crustaceans to the regeneration of the whole organism from one segment, as in certain annelid worms. Regeneration is common in plants and occurs naturally, as in VEGETATIVE PROPAGATION, or can be induced to propagate plants of economic importance, e.g. potato and tobacco plants. Complete regeneration of any plant is only possible if its vegetative cells have retained the full genetic potential (i.e. are TOTIPOTENT), enabling them to replicate every part of the plant.

regression the tendency to return to an average state from an extreme one.

reinforcement describes a certain situation in animal behaviour. A particular behavioural act is made more or less likely to occur, or the intensity of it is strengthened or lessened, by presenting a stimulus (the reinforcer) just after the act has taken place. A positive reinforcer such as a reward enhances the behavioural act whereas a negative reinforcer such as a punishment has a dampening down effect. This is used in the process known as *conditioning*.

relative atomic mass (*formerly called* **atomic weight**) the mass of atoms of an element given in atomic mass units (u), where $1u = 1.660 \times 10^{-27}$ kg.

releasing hormone one of several hormones synthesized by neurosecretory cells in the HYPOTHALAMUS of the brain in vertebrates. It is released in to the blood and either stimulates or inhibits the anterior PITUITARY GLAND to secrete its hormones.

renin a proteolytic ENZYME (that breaks down protein) that is produced by glomerular cells in the vertebrate KIDNEY under the control of the SYMPATHETIC NERVOUS SYSTEM. When blood pressure drops and sodium levels fall, renin cleaves the protein angiotensinogen, which is present as a plasma protein, to form angiotensin I. This itself is converted in the lung to angiotensin II, which is a potent vasoconstrictor, i.e. it causes blood vessels to constrict. It raises blood pressure and causes the kidneys to retain sodium and excrete potassium and so is involved in OSMOREGULATION.

rennin an ENZYME, secreted by cells in the stomach wall of mammals, that causes milk to coagulate and is especially important in young animals. It acts on a soluble protein present in the milk, caseinogen, and converts it to casein, which is an insoluble form.

Casein remains in the stomach long enough to be broken down by protein-digesting enzymes, which would not otherwise be the case.

replication the duplication of genetic material, generally before cell division.

Reptilia the reptiles, a class of vertebrates which were the first animals to truly colonize dry land and to exist entirely independently of water. They now often inhabit extremely hot conditions such as semi-desert. They have a dry skin covered with horny (keratinized) scales, which protects them from desiccation, and they are true air breathers with trachea and bronchial tubes leading to the lungs. They do not have a DIAPHRAGM and air is drawn in and out by respiratory movements involving the ribs. They have a four-chambered heart, but the ventricles are usually not completely separate and there is a single aortic arch (*see* AORTA) which is an ancestral feature. The excretory system is similar to that of mammals and birds and is able to produce a HYPERTONIC urine which conserves water. The movement to land was made possible by the development of a shelled amniote egg. Eggs are usually laid on land and the protective shell is porous and allows for some exchange of gases. Fertilization is internal in reptiles, and eggs may be retained within the body until they hatch (*see* OVOVIVIPAROUS). Reptiles are cold-blooded (*see* POIKILOTHERMY) but are able to maintain an even body temperature by behavioural means, by basking in the sun. There is evidence that some fossil reptiles were warm-blooded. Modern reptiles include snakes, lizards, turtles, and crocodiles.

respiration the process by which living cells of an

organism release energy by breaking complex organic compounds into simpler ones using enzymes. Respiration can occur in the presence of oxygen (*see* AEROBIC RESPIRATION) or in its absence (*see* ANAEROBIC RESPIRATION) and has an initial stage called GLYCOLYSIS, which is common to both forms of respiration.

The term respiration is also used, although less frequently, for gaseous exchange (better known as breathing) in an organism, which involves the uptake of oxygen from, and the release of carbon dioxide to, its surrounding environment.

respiratory chain *see* **electron transport chain**.

respiratory organ any organ present in an animal through which the gaseous exchange of oxygen and carbon dioxide takes place. It characteristically has a thin membrane with a good supply of blood, and examples are the TRACHEAE of arthropods, GILLS and gill books and vertebrate lungs.

respiratory pigment a substance that may be coloured and is found in animal tissues or blood and which transports oxygen from respiratory organs to body tissues. It forms weak bonds with oxygen, binding it when concentrations are high and releasing it when they are low. Some give colour to the blood, such as HAEMOGLOBIN, which is the respiratory pigment contained in the red blood cells (erythrocytes) of vertebrates.

resting potential the difference in electrical potential across the membrane of a nerve cell when an IMPULSE is not being transmitted. The inside is negative with respect to the outside and this is maintained by transport proteins, namely the SODIUM-POTASSIUM PUMP and the *potassium leak channel*. The sodium pump uses ATP to pump sodium out and potassium in. The potassium leak channel allows potassium ions to flow out along their electrochemical gradient. Eventually an equilibrium potential for potassium ions is reached, which in nerve AXONS is -75mV (*see* ACTION POTENTIAL).

restriction enzyme a type of enzyme found in bacteria which cleaves foreign DNA (i.e. that of an invading VIRUS) while leaving the host cell DNA unharmed. During replication of the bacterial DNA, the bases are altered and this protects it from attack by its own cleavage enzymes. Restriction enzymes destroy the DNA at particular sites along the strand. They are named and numbered according to the host organism in which they are found and the point at which they cleave the DNA. They are widely used in GENETIC ENGINEERING techniques.

restriction mapping a method of identifying the sites at which a DNA strand is cleaved by restriction enzymes. This is used in various genetic techniques such as CHROMOSOME mapping, and is of value in detecting genetic abnormalities in a foetus.

reticular fibres fine, branched protein fibres composed of reticulin that form a tightly woven mesh in many vertebrate CONNECTIVE TISSUES. They help to bind tissues or organs together and are found, for example, around muscle and in lymphoid tissue. The protein is produced by cells known as fibroblasts and the network is extracellular.

reticular formation a number of nerve cells within the BRAIN STEM of vertebrates, which may be congregated together to form nuclei. The reticular formation receives information from and sends impulses to the CEREBRAL HEMISPHERES, CEREBELLUM and SPINAL CORD. It regulates states of arousal and is involved in waking and alertness, and also consciousness in higher vertebrates.

reticulocyte an immature red blood cell in mammals that is non-nucleated. It is formed from a myoblast cell and reticulocytes are present in the blood during active phases of HAEMOPOIESIS (blood formation). They characteristically occur in increased numbers in the blood of a person with haemolytic anaemia.

reticulum (*plural* **reticula**) found in ruminant mammals, this is the second compartment of the stomach, which receives partially digested food from the rumen. This food has been rechewed and swallowed a second time and is subsequently passed on to the abomassum, the third compartment of the four-chambered stomach. Water is pressed out of the food in the reticulum.

retina the light-sensitive layer that lines the interior of the vertebrate (and cephalopod) eye and contains the photosensitive cells, the RODS and CONES. The retina is composed of two layers, the outer one being pigmented to prevent light from being reflected back. The inner layer contains the rods and cones and also blood vessels and nerve cells. Rods and cones are stimulated by light and synapse with intermediary NEURONS, which transmit impulses to the brain via the optic nerve (*see also* EYE, and OPTIC NERVE).

retinol *see* **vitamin A**.

retrovirus a VIRUS that attacks higher animals such as mammals and contains RNA. The RNA is converted to DNA by an enzyme called REVERSE TRANSCRIPTASE, which enables it to become inserted into the DNA of the host cells. These viruses are ONCO-

GENIC and contain cancer-causing oncogenes. The genes become active when the virus starts to replicate within the host cell.

reverse transcriptase a DNA POLYMERASE ENZYME found in RETROVIRUSes. It synthesizes the formation of a complementary strand of DNA on the single strand RNA template of the virus. Hence the GENOME (genetic material) of the virus becomes incorporated into the DNA of the host and is replicated. This enzyme has important uses in GENETIC ENGINEERING techniques.

rhesus *see* **blood grouping**.

rhizome a horizontal underground stem that bears axillary buds in papery, scale leaves. It is the means by which the plant survives the winter and may be a means of vegetative propagation. It has a root-like appearance, and examples are the nettle, couch grass and iris.

rhodopsin *also known as* **visual purple** this is the light-sensitive pigment found in the RODS of the RETINA of vertebrates. It contains a protein component, opsin. When light is received by the rod cell, the rhodopsin undergoes a chemical change and this causes a nerve impulse to be produced. Rhodopsin is highly sensitive and allows for vision in dim light.

rhombencephalon *see* **hind brain**.

rib one of a number of thin, paired, skeletal bones in the thorax of vertebrates that together form the rib cage. This protects major organs such as the lungs and heart. The ribs articulate at the back with the thoracic VERTEBRAe of the spinal column. In mammals, birds and reptiles the ribs articulate at the front with the breast bone or sternum. Their movement is controlled by intercostal muscles and this is important in the VENTILATION of the lungs.

riboflavin one of the VITAMINs of the B complex, B_2, which is widely available in foods such as milk, yeast and liver. It is a precursor of two coenzymes, FMN and FAD, which are involved in the OXIDATIVE PHOSPHORYLATION reactions of the ELECTRON TRANSPORT CHAIN and in energy metabolism. If deficient in the diet, it causes sores around the mouth and eyes in humans.

ribonucleic acid *see* **RNA**.

ribose a PENTOSE, MONOSACCHARIDE carbohydrate with 5 carbon atoms in its structure ($C_5H_{10}O_5$), which is important as a constituent of RNA. Deoxyribose is derived from it ($C_5H_{10}O_4$), and this is also very important, being a component of DNA. It is usually combined in another structure and is seldom found in its free form in living organisms.

ribosomal RNA (rRNA) one of the three major classes of RNA, which is transcribed from DNA in a structure of eucaryotic nuclei called the NUCLEOLUS. Along with many PROTEINS, ribosomal RNA forms the cellular structures called RIBOSOMES, which are found in both EUCARYOTIC and PROCARYOTIC cells.

ribosome the cellular structure that is the site of PROTEIN synthesis in all EUCARYOTIC and PROCARYOTIC cells. Ribosomes are composed of one large and one small sub-unit, which contain RIBOSOMAL RNA and associated proteins. Analysis of procaryotic and eucaryotic ribosomes indicates that they share the same evolutionary origins as their structure, and the RNA they contain (except a segment unique to eucaryotes) are virtually identical. Ribosomes assemble at one end of a MESSENGER RNA molecule and move along the molecule to build the POLYPEPTIDE chains of all proteins in the TRANSLATION process.

ring compound *see* **closed-chain**.

Ringer's solution a physiological saline that is used to nurture cells or organs in the laboratory. It contains calcium, potassium and sodium chlorides in solution, buffered either with phosphate or bicarbonate.

RNA (*abbreviation for* ribonucleic acid) a complex NUCLEIC ACID, widely found in living cells, and usually consisting of a single strand of NUCLEOTIDES. The nucleotides are composed of ribose sugar and the organic bases ADENINE, CYTOSINE, GUANINE and URACIL combined with phosphate. RNA is involved in PROTEIN SYNTHESIS, and in some viruses it is also the genetic material and may be double stranded (*see* RETROVIRUS and REVERSE TRANSCRIPTASE). RNA is mostly produced in the nucleus and then distributed within the CYTOPLASM of the cell. There are three kinds, MESSENGER RNA (mRNA), RIBOSOMAL RNA (rRNA) and TRANSFER RNA (tRNA).

rod cell a highly sensitive light receptor cell present in the vertebrate RETINA and containing the pigment RHODOPSIN. They allow vision in very dim light and are unevenly distributed within the retina, being absent from the fovea (the area of most acute vision) but concentrated at the lateral margins. Rods do not detect colour, and they are the commonest light receptors present in the human retina. There are about 125 million, compared to 6 million CONE cells. Most mammals are nocturnal and have poor colour vision, but keen night sight is conferred by the large number of rods present.

Rodentia a widespread and successful order of mammals containing the rats, mice, squirrels, capybaras and beavers. The upper and lower jaws contain a single pair of long INCISOR teeth, which grow continuously throughout the animal's life. These are adapted for gnawing, and the absence of enamel on the back means that they wear to a chisel-like cutting edge. Rodents are herbivorous or omnivorous and tend to breed prolifically. Some are serious pests, spoiling food and are implicated in the spread of disease, notably rats.

röntgen (R) a radiological term defining the X-ray or gamma ray dose producing ions carrying a specific charge.

röntgen equivalent man a unit of radiation dose that has now been replaced by the SIEVERT.

root a plant organ that tends to grow downwards, penetrating the soil surface in response to gravity, anchoring the plant, and also providing it with nutrients and water. It differs from a stem in having the vascular tissue (XYLEM and PHLOEM) within a central core and never containing CHLOROPHYLL. In addition, roots do not bear buds, leaves or flowers but do possess a ROOT CAP. The embryonic root, or radicle, usually develops either a fibrous root system, with many roots of similar size, or a tap root system. In the latter there is one main tap root from which smaller lateral roots develop. Root hairs develop from the EPIDERMIS just behind the area of active growth, the root tip. These are in intimate contact with the soil and increase the surface area for absorption of water and minerals. They are continually being replaced by the new root tissue arising from the apical MERISTEM at the root tip. The apical meristem is the region of actively dividing cells from which all the primary root tissues arise.

root cap a protective cap or sheath of cells over the apex of a ROOT that protects the root as it pushes through the soil.

root hair also known as a *pilus*, a fine tube projecting from an epidermal ROOT cell. Root hairs form a piliferous layer and increase the available surface area for the absorption of water and nutrients.

rRNA *abbreviation for* RIBOSOMAL RNA.

ruminant any mammal that has four compartments in the stomach to aid the digestion of large amounts of plant matter. Ruminants (order Artiodactyla) include cattle, sheep, deer and giraffes. In the first section of their stomach, the rumen, food is enveloped in a mucus and is partially digested by an ENZYME called cellulase, which is supplied by the billions of bacteria living in the rumen. After this, the food is regurgitated to the mouth, and, after chewing, it passes through a further two sections (water is removed) and eventually ends up in the true stomach, the abomassum, which contains the enzymes essential for complete digestion.

S

saccharides SUGARS (and therefore CARBOHYDRATES) divided into mono-, di-, tri- and polysaccharides. Monosaccharides are the basic units, simple sugars; disaccharides, e.g. sucrose and lactose, are formed by condensing two monosaccharides and removing water. Sucrose gives, on HYDROLYSIS, a mixture of glucose and fructose; trisaccharides comprise three basic units and polysaccharides are a large class of natural carbohydrates including STARCH and CELLULOSE.

saccharin a white, crystalline powder with about 500 times the sweetening power of sucrose. It is not very soluble in water but is used extensively as a sweetening agent, in the form of the sodium salt.

Saccharomyces a genus of YEASTS that are widely used in the production of bread and alcoholic beverages.

sacral vertebrae the VERTEBRAe that occur in the lower back lying between the lumbar and caudal vertebrae. They articulate with the PELVIC GIRDLE and are frequently fused to form a single bone known as the SACRUM. This provides firm support for the pelvic girdle. Mammals have at least three sacral vertebrae, reptiles two and amphibians only one.

sacrum (*plural* **sacra**) a group of SACRAL VERTEBRAE that are fused together.

saline a solution containing a salt (see also RINGER'S SOLUTION).

saliva a watery secretion produced by the salivary glands in the mouth or by the labial glands of insects. In higher animals, the sight, smell or presence of food in the mouth, or even the thought of food, results in the secretion of saliva. It contains mucus which lubricates the food and makes it easier to swallow. In many terrestrial vertebrates, amylase enzymes (such as PTYALIN in mammals) are present in saliva and begin the breakdown of starch in food. Many digestive enzymes are found in the saliva of insects, while that of blood-sucking animals such as some insects, leeches and vampire bats contains anticoagulants. PATHOGENIC organisms may also be passed on in this way, in saliva.

Salmonella a genus of bacteria that are gram-negative (*see* GRAM'S STAIN) and rod-shaped. They are important because they contain species that cause serious diseases in man and animals. They are generally motile and AEROBIC and occur in the intestine but can exist for a long time in water and sewage. Infection occurs as a result of the consumption of food containing the bacteria, for example, eggs or meat. Alternatively the food may be subsequently contaminated by a human or animal carrier. Various species of Salmonella cause diseases such as salmonellosis, gastroenteritis, septicaemia and typhoid fever.

salt a compound formed when a metal ATOM replaces one or more hydrogen atoms of an ACID (*see also* BASE).

saprotroph an organism that obtains its nutrition from dead and decaying organic matter. The group includes many bacteria and fungi, which are responsible for the release of nitrogen, carbon dioxide and other nutrients from the decomposing matter.

sapwood lighter in colour than HEARTWOOD, it is the outer wood of branches and tree trunks, containing living cells, e.g. XYLEM, PARENCHYMA and medullary ray cells. Its main function is to conduct water and store food but it also provides support for the plant.

saturated compound a group of compounds with no double or triple BONDS, i.e. they do not form addition compounds through the joining of hydrogen atoms or their equivalent.

saturated solution a SOLUTION of a substance that exists in equilibrium with excess SOLUTE present.

savanna areas of grassland in tropical or sub-tropical zones, occupying a broad zone between the tropical forests and the semi-arid steppes. They are a result of prolonged lack of water in soils and usually occur in areas of low relief where rainfall is absent because of the presence of mountains.

scales small bony or horny plates covering the body surface of fish and reptiles and also present on the legs of birds and tails of some rodents. Scales are also found on the wings of the LEPIDOPTERA (butterflies and moths) and these are modified hairs arising from the CUTICLE, which often confer colour. Reptiles may have two types of scales present, the first being horny and epidermal in origin (corneoscutes). These are sometimes fused with scales lying deeper within the dermis which are called osteocutes. Three types of scales are present in fish and these are used as a means of classification. *Placoid* *scales* are characteristic of the cartilaginous fish (CHONDRICHTHYES) and these have an outer enamel layer and an inner core of DENTINE and each layer bears a projecting spine. Teeth are thought to be modified placoid scales. The coelacanths and lung-fish have characteristic cosmoid scales composed of an outer layer of cosmin (similar to dentine), which is covered by an enamel-like surface of ganoine. The inner layers of these scales are bony, and growth is achieved by adding to these. Some other primitive ray-finned fishes (order Acipenseriformes), such as the sturgeon, have similar *ganoid scales* in which growth is achieved by adding new material all round.

scanning electron microscope *see* **electron microscope**.

scapula (*plural* **scapulae**) the largest bone present in the vertebrate PECTORAL GIRDLE, which is the shoulder blade of mammals. It articulates with the HUMERUS and CLAVICLE and anchors muscles operating the forelimb.

Schiff's reagent devised by the German chemist Hugo Schiff (1834-1915), it is a reagent used to test for the presence of aldehydes. It is a solution of fuschin dye but the colour has been taken out by the addition of sulphur dioxide. Aldehydes and some ketones restore the pink colour if they are present in the solution.

Schistosoma a blood fluke, parasitic in man, and belonging to the class TREMATODA, that causes the serious disease bilharzia, or schistosomiasis. The adult flukes are present in man in the blood vessels of the intestine and lay eggs which penetrate the walls causing severe damage. Fertilized eggs pass either into the bladder, and out with the urine, or into the rectum passing out with the faeces. The mode of exit depends upon the species of fluke. The larvae that eventually hatch are called miracidia. The miracidium burrows into a snail, which is the intermediate host, and there daughter sporocysts are formed, which emerge in water as another larval form, the cercaria. The cercaria burrow into the skin of humans often while they are planting rice crops or bathing. This is a serious disease affecting many people especially in China, South America and Egypt. Treatment includes the use of drugs and education so that people no longer enter the water unprotected. There are also attempts to control the intermediate snail host.

Schwann cell a cell named after the German anatomist T. Schwann (1810–82), which is specialized for the formation of the MYELIN SHEATH around a nerve AXON, known as an internode. Internodes are interrupted by non-myelinated parts, the NODES OF RANVIER. The myelin sheath comprises concentric layers of Schwann cell membrane wrapped around the axon.

scientific notation a useful method of writing large and small numbers. The scientific notation for a number is that number written as a power of 10 times another number, x, such that x is between 1 and 10 (1 < x <10), e.g. $145,800 = 1.458 \times 10^5$.

scion *see* **graft**

sclera the white membrane present in the vertebrate EYE, forming the tough exterior sclerotic layer. It is modified to form the CORNEA and is composed of fibrous tissue.

scleroproteins insoluble PROTEINS that form the

skeletal components of tissues. Included are KERA-TIN, COLLAGEN and elastin (a fibrous protein found in lungs, artery walls and ligaments).

scrotum a pouch of skin and tissue present in male mammals, which houses the TESTES and supports them. The scrotum is an external structure and the temperature within it is lower than that inside the body. This is necessary because sperm production is impaired if the temperature is raised.

Scyphozoa a class of the phylum CNIDARIA containing the jellyfish, which typically have the MEDUSA as the dominant body form and are marine. Many have a life cycle which exhibits an *alternation of generations* and they have an EPIDERMIS bearing FLAGELLAe and numerous cnidoblasts (stinging cells). They have distinct muscle cells, which are present in a layer beneath the epidermis and the mesogloea contains many cells and fibres. A coelenteron (a central gastrovascular cavity) is present, lined with gastric filaments and often divided into four gastric pouches arranged radially. The filaments secrete digestive ENZYMEs. The pouches are normally divided by septa but have complicated connecting canals. The Scyphozoa usually have four gonads located in the floor of the gastric pouches. The sexes are separate and fertilized eggs develop into planula larvae which are CILIAted. The planula settles on the sea floor and becomes another larval form, with tentacles, the scyphistoma. It may exist in this form for several years but ultimately becomes either a single larval medusa or forms several of these by a form of budding (*see* ASEXUAL REPRODUCTION) called strobilation. The larval medusae so produced are known as ephyrae (*singular* ephyra). Asexual reproduction occurs only at this stage, and adult jellyfish always reproduce by the formation of GAMETES.

seat earth a fossil soil (palaeosol) that is found immediately beneath coal seams and represents the soil in which the vegetation grew. It is thus the last sediment deposited before plant life became established and frequently contains fossilized roots.

sea water all but 0.1 per cent of material dissolved in sea water is due to eleven components (see below). The bulk of the calcium and bicarbonate ions precipitate out as calcium carbonate ($CaCO_3$), silica (SiO_2) is taken up by organisms and most of the material dissolved consists of five ions: chloride, sodium, sulphate, magnesium and potassium.

ion	‰
chloride Cl^-	19.0
sodium Na^+	10.6
sulphate SO_4^{2-}	2.6
magnesium Mg^{2+}	1.3
calcium Ca^{2+}	0.4
potassium K^+	0.4
bicarbonate HCO_3^-	0.1
bromide Br^-	}
borate H_3BO_3	} less than 0.1‰
strontium Sr^{2+}	}
fluoride F^-	}

sebaceous gland a small gland present in the skin of mammals that secretes an oily substance called sebum into the base of a HAIR FOLLICLE. Sebum is a fatty liquid that lubricates the fur, hair and skin, providing waterproofing and protection.

secondary metabolite a general term applied to several groups of compounds that are not directly involved with the biological processes that contribute to growth, such as photosynthesis and respiration. However, they may be chemicals used in defence or similar mechanisms. Typical groups are TERPENOIDS, alkaloids and the flavonoids (plant pigments).

secretin a POLYPEPTIDE HORMONE produced by cells lining the anterior SMALL INTESTINE (duodenum and jejunum) in vertebrates in response to the presence of acid from the stomach. It stimulates the production of pancreatic juice, which is alkaline, and also bile secretion in the liver, but it inhibits the release of gastric juices. Secretin was the first substance to be named as a hormone.

secretion the process by which particular substances or solutions are produced and discharged by specialized cells in living organisms. The material produced is also known as the secretion and the cells that produce it may be collected together within a GLAND.

sedimentary rock one of the three main rock types. Sedimentary rocks are formed from existing rocks through processes of erosion, denudation and subsequent deposition, compaction and cementation. The main types are: terrigenous—derived from existing rocks on the land, e.g. sandstones and shales; organic—produced by organic processes, e.g. limestones formed from coral reefs; chemical—precipitated from solution, e.g. evaporites such as gypsum; and volcanogenic—associated with volcanic action, e.g. volcanic ash deposited as tuff, or bentonite.

sedimentary structure a fossil feature, preserved in or on the bedding of a sedimentary rock. The structures are those generated by sedimentary processes or the activity of organisms at the time of deposition. Structures preserved on bedding surfaces include ripple marks, scour marks (caused by erosion) and tool marks (caused by an object being carried over the surface), and these are all formed by depositional processes. Sole marks are structures preserved on the bases of beds and include trails, tool marks and infilled scours and load casts (a bulging formed at the base of a bed where the upper bed sinks into the lower while the sediment is wet—forming a lobe). Internal sedimentary structures include those formed by depositional processes, e.g. lamination, convolute bedding (because of expulsion of water from sediments deposited quickly) and slump structures (overturned folds caused by sliding sediment); organic activity, e.g. bioturbation, or by chemical activity after deposition, e.g. concretions.

seed the product of fertilization in angiosperm and gymnosperm plants (*see* ANGIOSPERMAE and GYMNOSPERMAE), which develops from the OVULE. It consists of an embryo and tissue that supplies nutrients. This is either endosperm, which surrounds the seed or, in non-endospermic plants, food may be stored in the *cotyledons*. The seed is enclosed by a protective seed coat, the testa, which is formed from the integuments. In angiosperms, the OVARY wall develops into a fruit that encloses the seeds. Gymnosperms do not produce a fruit and their seeds are described as naked. The development of seeds represented an evolutionary advance for plants, enabling them to colonize the land, as they were no longer dependent upon water for fertilization. Seeds provide protection and nutrition for the embryonic plants and can be dispersed by the wind or animals. Seeds are also a means by which the plant can survive the winter or other harsh conditions until the environment is favourable for germination.

segmentation *see* **metamerism**.

semen the fluid that contains the sperm that is ejaculated by a male mammal during copulation. The sperm (SPERMATOZOA) are produced by the TESTES and the nutrient liquid is secreted by the seminal vesicles, cowper's gland and PROSTATE GLAND.

semicircular canals a structure in the INNER EAR in vertebrates that helps maintain dynamic equilibrium (balance). It comprises three "loops" mutually at right angles, and in each loop is a fluid (the endolymph). When the head is moved, the fluid moves accordingly and sensory cells respond to the movement of the endolymph. The nerve impulses are then transmitted to the brain.

seminal vesicle an organ present in some invertebrates such as insects and also in lower vertebrates. It is a sac that stores sperm. In male mammals, it is one of a pair of glands arising from the VAS DEFERENS (sperm duct) that contributes an alkaline liquid secretion to the semen.

senescence the process of ageing that is characterized by the progressive deterioration of tissues and the metabolic functions of their cells. According to research, senescence may be caused by the accumulation of genetic mutations within the body's cells or the expression of undesirable GENES in the later years of an individual's life. Some organisms are able to suppress senescence by REGENERATION, a process common in many simple invertebrates and one achieved by some plants using VEGETATIVE PROPAGATION.

sense organ an organ present in an animal's body that contains a concentration of RECEPTORS, specialized to be sensitive to a particular type of stimulus. Examples are the ears, eyes and nose.

sepal a part of a flower present in angiosperms (*see* ANGIOSPERMAE). A number of sepals make up the calyx, which is an outer supporting structure for the rest of the flower. Sepals are usually green, resembling leaves, and often bear hairs. However, sometimes they can be coloured and more like petals. They are thought to be modified simple leaves.

septum (*plural* **septa**) any dividing wall or partition found in plants and animals, which may be complete or perforated. Examples are the intersegmented septa of earthworms and the septa that separate the chambers of the vertebrate heart.

serology the study of blood serum within the laboratory, particularly ANTIGEN-ANTIBODY reactions. The study of these is important in order to gain insight into the mechanisms of the IMMUNE SYSTEM.

serous membrane a membrane that lines an internal body space such as the pleural, peritoneal and pericardial cavities in mammals. It consists of a layer of MESOTHELIUM overlying a thin layer of CONNECTIVE TISSUE that attaches it to the lining of the body cavity.

serum *see* **blood serum**.

sessile a term usually applied to animals that live permanently attached to a surface and are sedentary, such as many marine organisms, e.g. barnacles, sponges and corals. It also describes a structure that would be expected to have a stalk but where this is, in fact, absent. Examples are oak (*Quercus robur*) leaves, which are attached directly to twigs, and also the eyes of some crustaceans.

seta (*plural* **setae**) a fine hair or bristle present in many invertebrates and arising from the epidermis. It is either a hollow projection arising from the CUTICLE and containing part of or a whole epidermal cell (e.g. insects), or it is composed entirely of CHITIN and projects from the cuticle as in a chaeta of an annelid worm. In some plants, a seta is a thin stalk arising from the SPORANGIUM, as in the liverworts and mosses. In ALGAE a seta is a hollow extension of a cell forming a stiff hair-like structure.

sex chromosome one of a pair of chromosomes that play a major role in determining the sex of the bearer, with a different combination in either sex. An individual is said to be homogametic when it has a HOMOLOGOUS pair of sex CHROMOSOMES (as in the XX of female mammals) and is said to be heterogametic when it has different sex chromosomes forming its pair (as in the XY of male mammals). Sex chromosomes contain GENES that decide an individual's sex by controlling the sexual characteristics of the individual, e.g. testes in human males and ovaries, breasts, etc, in human females.

sex linkage the location of a GENE on a SEX CHROMOSOME, although the expression of the gene does not necessarily affect the sexual characteristics of the individual. Some examples of sex-linked genes include red-green colour blindness and haemophilia, both RECESSIVE genes found on the X-chromosomes. Such X-linked genes cannot be passed from father to son, as the father contributes only the Y-chromosome while the mother contributes the X-chromosome to a son. The Y-chromosome contains fewer specific genes than the X-chromosome other than those responsible for maleness, and any Y-linked genes will only be inherited by male offspring.

sexual reproduction the production of progeny that have initially arisen from the fusion of male and female GAMETES in a process called FERTILIZATION. In DIPLOID organisms, sexual reproduction must be preceded by MEIOSIS to form the HAPLOID gametes if there is not to be a doubling of the number of chromosomes in all sexually reproduced offspring. As sexual reproduction involves MEIOSIS, it introduces greater genetic variation in a species, because GENETIC RECOMBINATION can occur during meiosis, with the result that any offspring will have gene combinations that differ from its parents.

shield a screen placed around persons or equipment to offer protection against harmful rays (e.g. X-rays) or to protect an electronic component from interference from electromagnetic fields.

short-sightedness *see* **myopia**.

shoulder girdle *see* **pectoral girdle**.

sickle-cell anaemia a type of inherited, genetically determined haemolytic anaemia that affects African people and African-Americans. It is caused by a recessive ALLELE, i.e. to exhibit the disease a person must be HOMOZYGOUS for the sickle-cell trait. A single AMINO ACID in the protein chain of a normal HAEMOGLOBIN molecule is substituted and this causes the disease, which is characterized by the red blood cells deforming to a sickle shape. This causes a number of symptoms including blood clotting. Many people (about one in ten African-Americans) are HETEROZYGOUS for the sickle-cell trait and are carriers for the disease but remain healthy themselves. The sickle-cell allele would appear to be selected for, and thus remains at a high level in, African populations because in heterozygotes it confers increased resistance to the most serious form of malaria.

sievert (Sv) the SI unit of radiation dose defined as that radiation delivered in one hour at a distance of one centimetre from a point source of one milligram of radium enclosed in platinum that is 0.5mm thick.

significant figures the digits in a number that contribute to its value, e.g. in the number 0.762 the zero is insignificant whereas the other digits are significant. If the number were 0.7620, the last zero ought to be significant because it should indicate that the number is accurate to four decimal places. However, the final zero as shown here is often added arbitrarily by the originator of the data and it may not represent such accuracy.

silica gel the hard amorphous form of hydrated silica, which is chemically inert but highly hygroscopic (a substance that absorbs moisture) It is used for absorbing water and solvent vapours and other drying or refining tasks and can be regenerated by heating.

silicula *see* **siliqua**

siliqua a special type of capsule formed in plants such as wall flowers. It is a dry fruit formed from an ovary of two fused CARPELs that are divided by a central SEPTUM into two compartments, the locules. The fruit so formed is longer than it is broad and when it dries out the two compartments separate to expose and liberate the seeds. A *silicula* is a very similar fruit but it is short and broad and is found in such plants as candytuft and *Alyssum*.

Silurian a geological period of the Palaeozoic era, following the ORDOVICIAN and preceding the DEVONIAN.

It lasted from 438 to 408 million years ago and was named by Roderick Murchison (1792-1871) who first observed Silurian rocks in Wales. He named the rocks after an ancient tribe of the region, the Silures. Most organisms were marine and common fossils are brachiopods (*see* BRACHIOPODA), corals and crinoids. GRAPTOLITES and TRILOBITES declined but are still seen as fossils in Silurian rocks. Primitive jawless fish are the only vertebrate fossils found (the Ostracoderms, now extinct) and jawed fish made an appearance later in the period. The phase of mountain building known as the Caledonian orogeny was at its greatest towards the end of the Silurian.

sinus a sac-like depression or cavity present in an animal, e.g. the nasal sinuses present in the bones of the face in mammals.

sinusoid a type of minute blood vessel or small space filled with blood within an organ of an animal. In some organs such as the liver, sinusoids replace capillaries, allowing immediate contact between the tissue and its blood supply.

SI units a system of coherent metric units—Système Internationale d'Unités (*see* APPENDIX 4).

skeleton the whole structure that provides a framework and protection for an animal's body and within which organs are protected and muscles are anchored. There may be an EXOSKELETON as in many invertebrates or an internal ENDOSKELETON both of which have flexibility conferred by JOINTS (*see also* BONE).

skin an organ comprising the outer layer of the body of a vertebrate animal. There is an external EPIDERMIS itself consisting of an outer STRATUM CORNEUM beneath which lies a granular layer with a Malpighian layer at the base. The Malpighian layer is the region of growth that produces the epidermis. Beneath the epidermis is the DERMIS and a layer of subcutaneous tissue composed mainly of fat. The subcutaneous tissue contains glands of various types (e.g. SEBACEOUS, sweat and mucus glands), sensory receptors sensitive to temperature, pain and pressure, and nerves, muscles and blood capillaries. The skin may bear a variety of structures protruding from it—commonly hair, fur, feathers and scales. It forms a protective layer for the body helping to cushion it in the event of mechanical injury, preventing desiccation and providing insulation. It also allows for some exchange of gases in amphibians.

skull the part of the vertebrate skeleton that forms the head and encloses the brain. In mammals it consists of a number of bones that are separated by (mainly) immovable joints. There is a cranium enclosing the brain and other facial and jaw bones. The only movable joint is that which articulates the lower mandible (jaw bone) to the rest of the skull. A large opening at the base of the skull called the FORAMEN magnum allows the spinal cord to pass from the brain to the trunk of the body.

sleep a state in animals of lower awareness accompanied by a reduction in metabolic activity and general physical relaxation. It is exhibited by many animals and is easily and often quickly reversible. When a person is about to fall asleep there is a change in the brain's electrical activity, which can be recorded by an ELECTROENCEPHALOGRAM. There are high amplitude, low frequency waves (slow wave sleep) interrupted by short phases of low amplitude, high frequency waves. During these phases the person may be restless, dream and exhibit rapid eye movements and this is known as REM sleep. The RETICULAR FORMATION of the brain is especially involved in the control of sleep although other areas are also active.

slime moulds simple organisms that are generally placed in the kingdom PROTISTA although their taxonomic position is uncertain. They resemble fungi in appearance but in the detailed life cycle show greater similarities to amoeboid organisms. They may be unicellular or form multicellular aggregates and occupy damp and aquatic habitats. Four phyla are recognized: the Myxomycota (plasmodial slime moulds), Acrasiomycota (cellular slime moulds), Oomycota (water slime moulds) and Chytridiomycota (saprobes that obtain nourishment from dead organic matter, or parasites).

small intestine a part of the vertebrate ALIMENTARY CANAL leading from the stomach to the large intestine in which there are three areas: the DUODENUM, JEJUNUM and ILEUM.

smooth muscle an involuntary muscle that is not under conscious control but is regulated by the AUTONOMIC NERVOUS SYSTEM in vertebrates. It is characterized by elongated cells that are spindle-shaped and lack striations. They may occur singly or in bundles, or sheets and the position of the nucleus varies between the different muscle cells. Smooth muscle is found, for example, in the gut, respiratory tract, around HAIR FOLLICLES and in the circulatory system. Contraction may be regulated by a built-in PACEMAKER.

soap the sodium or potassium salts of the FATTY ACIDS, stearic, palmitic and oleic acid. Soaps are produced by the action of sodium or potassium hydroxide on fats.

sodium (Na) an alkali metal, which does not occur in the free state naturally because it is highly reactive. Its principal source is salt (NaCl) occurring in sea water and salt deposits and the metal is obtained by ELECTROLYSIS of fused NaCl. The elemental metal is silvery-white, soft and can be cut with a knife. It reacts violently with water, and rapidly with oxygen and halogens (chlorine, fluorine, bromine, etc). Sodium is an essential element for animal cells. It is vital in the control of the pH of body fluids and in the transmission of nerve impulses (*see* ACTION POTENTIAL, SODIUM-POTASSIUM PUMP).

sodium hydroxide caustic soda, NaOH. A whitish, deliquescent (*see* DELIQUESCENCE) substance that gives a strongly alkaline solution in water. It is used a great deal in the laboratory and is a very important industrial chemical. It is used in the manufacture of soap, paper, aluminium, petrochemicals and many other chemicals and products.

sodium-potassium pump a form of active transport requiring ATP as its energy source, by which sodium ions are pumped out of a NEURON through the cell membrane. In addition, potassium ions are pumped in, in the ratio of three sodium to two potassium ions. This mechanism maintains and regulates the RESTING POTENTIAL of the cell membrane. The sodium-potassium pump is an electrogenic pump that stores energy in the form of voltage and is the main such system operating in animal cells.

solute one substance dissolved in another. A solute dissolves in a SOLVENT to form a SOLUTION.

solution a single phase mixture of two or more components, which usually applies to solids in liquids and often refers to a solution in water (aqueous solution). However, other solutions include gases in liquids and liquids in liquids.

solvent a substance, usually a liquid, that can dissolve or form a SOLUTION with another substance.

somatic cell any of the cells of a multicellular organism (plant or animal) other than the reproductive cells (GAMETES).

somatotrophin *see* **growth hormone**

species a group of individuals that can potentially or actually interbreed producing viable offspring, and that within the group may show gradual morphological variations but remain different to other groups. In the taxonomic classification, species are grouped into a genus (plural *genera*) and species can themselves be subdivided into subspecies, varieties, etc. The naming of species, etc, is governed by the system of *binomial nomenclature*, so that a generic and specific name (given in Latin) identify a particular individual, e.g. *panthera pardus* is the leopard.

spermatheca a small sac-like receptacle present in some hermaphrodite and female animals, in which sperm is received from the mating partner and stored until the eggs are ready for fertilization. It is found, for example, in earthworms.

spermatogenesis the process by which spermatozoa (sperm) are produced by a series of cell divisions within the testes. The TESTIS contains seminiferous tubules within which germ cells undergo MITOSIS to produce spermatogonia. These divide further by mitosis to produce primary spermatocytes. This is followed by two phases of division by MEIOSIS producing, initially, secondary spermatocytes and then spermatids in which the chromosome number has been halved. The spermatids undergo further development to become mature spermatozoa or sperm cells. The whole process of spermatogenesis takes 65 to 75 days in human beings (*see also* OOGENESIS).

spermatophore produced by a number of animals that have internal fertilization, this is a small compact packet of sperm. Depending upon the species, the spermatophore is transferred into the female's body in a number of ways. In octopuses and squids the tentacles are employed to transfer spermatophores. In other animals, e.g. some salamanders, the spermatophore is dropped on the ground and picked up by the cloacal lips of the female as she passes over. Yet others use hypodermic-like structures to transfer the sperm through the skin of the female. Spermatophores are also produced by such animals as scorpions, spiders, mites, leeches and snails.

Spermatophyta the division of the plant kingdom containing all the plants that reproduce by the production of seeds. It is subdivided into the ANGIOSPERMAE and GYMNOSPERMAE.

spermatozoon the mature reproductive cell or gamete produced by male animals (*see* SPERMATOGENESIS). It has a head in which the HAPLOID nucleus with half the CHROMOSOME number is situated and there is also a structure called an acrosome. This is an organelle that helps the sperm to penetrate an egg. Behind the head there is a midpiece or middle section in which MITOCHONDRIA occur, to provide the energy necessary for the movement of the spermatozoon. Posteriorly there is a long thin tail consisting of a plasma membrane around a FLAGELLUM, which lashes from side to side to produce a forward movement.

sphincter a specialized ring of circular muscle that surrounds a tubular organ or controls the mouth of an orifice. A sphincter muscle contracts to close the orifice or tube and relaxes to open it. Examples are the pyloric sphincter at the lower end of the STOMACH and the anal sphincter at the ANUS.

spinal cord the part of the vertebrate CENTRAL NERVOUS SYSTEM that runs from the brain through the vertebral column, within the vertebral canal. Both GREY and WHITE MATTER are present, the former having an H-shaped cross-section with a hollow core. The cerebro-spinal canal, containing cerebro-spinal fluid, runs through the cross bar of the H. The grey matter is surrounded by white matter and the whole is covered by the MENINGES. Both sensory and motor NEURONS are carried by the spinal cord, and paired SPINAL NERVES arise from it, which leave through spaces between the vertebrae.

spinal nerves forming part of the PERIPHERAL NERVOUS SYSTEM, these are paired nerves that arise from each side of the SPINAL CORD, usually one pair leaving between each vertebra. They are mixed nerves containing both sensory and motor fibres. Each has a dorsal and ventral root, which usually fuse just outside the vertebral column. There are 31 pairs of spinal nerves in man.

spinneret a modified appendage that forms a small tubular organ that is used for the production of silk in spiders and some insects. The spinnerets of spiders occur on the posterior part of the abdomen and are thought to be modified legs. Silk is produced for the construction of webs, egg cocoons and also as cords to wrap around prey. The silk varies according to the use. It is secreted as a fluid, which then hardens when it is exposed to air. Insect spinnerets are not the same as those of spiders. Those of the silkworm are located in the pharynx, the silk being secreted by modified salivary glands.

spiracle (1) present on either side of the head in cartilaginous fish, this is one of a pair of small openings comprising the first GILL SLIT, which is very much reduced in size. Adaptations in the skeleton involving the mandibular and hyoid arches, which are used for firm attachment of the jaws, has resulted in the small size of the first gill slit (now spiracle) which is located between these two arches. In most living species of bony fish, the spiracles are entirely closed over. The analagous gill slit of embryonic tetrapod animals becomes the MIDDLE EAR cavity and EUSTACHIAN TUBE in adult animals. (2) In insects, the spiracles are the external openings of the respiratory TRACHEAE, which are often controlled by valves and are located along each side of the body.

spirillum any BACTERIUM with a spiral shape, which may be long and coiled and is also rigid. They are often highly mobile and may have tufts of FLAGELLA. They occur in water and soil, feeding on organic material.

spirochaete any BACTERIUM that has an elongated shape, twisted into a corkscrew, which is non-rigid, with a thin cell wall (*compare* SPIRILLUM). Some spirochaetes are free-living but others are parasitic and pathogenic, e.g. the organism that causes syphilis (*Treponema*). Spirochaetes are common in water polluted by sewage. They move by means of muscular contraction resulting in a wave-like movement through the cell.

spleen an organ present in vertebrate animals that comprises a mass of lymphoid tissue located behind the stomach. It is a site of formation of LYMPHOCYTES and acts as a reservoir for these and for blood platelets. In addition, it functions as part of the *reticulo-endothelial system*, where worn out red blood cells (and platelets) are destroyed and their iron is conserved. Erythrocytes (red blood cells) are stored in the spleen and their release regulated under the control of the SYMPATHETIC NERVOUS SYSTEM. Smooth muscle contraction releases the erythrocytes into the circulation, and at a time when the body is working hard may cause a "stitch" in man.

spontaneous generation a now discredited theory that living organisms could arise from non-life. It was believed that spontaneous generation could occur in, for example, rotting meat or fermenting broth, giving rise to an individual organism, but it is now known that all new organisms originate from the parent organism from whom they have inherited a genetic ancestry.

sporangium (*plural* **sporangia**) typically present in plants and fungi, this is the reproductive organ that produces asexual spores.

spore a small reproductive unit, usually consisting of one cell, that detaches from the parent and disperses to give rise to a new individual under favourable environmental conditions. Spores are particularly common in fungi and bacteria but also occur in all groups of green land plants such as ferns, horsetails and mosses.

sporophore present in certain fungi, this is a general term describing the spore-producing part. An example is the stalk and cap of a mushroom. Sporophores are aerial structures.

sporophyll a type of leaf present in plants on which the SPORANGIA are borne. It may resemble a foliage leaf and only be different because of the presence of the sporangia, as in bracken. Often, however, sporophylls are very specialized structures bearing no resemblance to ordinary leaves, e.g. CARPELS and STAMENS in flowering plants.

sporophyte the phase in a plant's life cycle that produces the spores. The sporophyte is DIPLOID but it produces HAPLOID spores. It may be the dominant stage in the plant's life cycle (as in the seed plants) or be mainly dependent upon the GAMETOPHYTE as in the bryophytes (*see* BRYOPHYTA).

Sporozoa a class of protozoan organisms now often classified as the phylum Apicomplexa. These are parasitic organisms of higher animals, some of which may cause serious diseases in man. They have a complex life cycle with sexual and asexual stages involving two or more different hosts. One of the most serious is the causal organism of malaria, *Plasmodium* (*see also* PARASITE).

Squamata an order of the class Reptilia comprising the lizards and snakes. They are covered by horny, epidermal scales and males uniquely have a paired penis. The lizards have movable eyelids and generally possess external ear openings. Other features include articulated halves of the lower jaw, a pectoral girdle, four limbs and a urinary bladder. Examples include iguanas, monitor lizards, chameleons and gila monsters. Snakes do not have external ear openings or movable eyelids, the eyes being covered by transparent eyelids. They lack a PECTORAL GIRDLE and limbs. The halves of the lower jaw are not articulated being attached only by ligaments, muscles and skin. This means that they can easily "dislocate" the jaw and swallow very large prey. Snakes do not possess a urinary bladder. Some are constrictor types, e.g. boas, pythons and anacondas, while others may be venomous, e.g. cobras, vipers, rattlesnakes, mambas and sea snakes.

staining a laboratory technique in which biological material that would normally be transparent and difficult to see under a microscope, is induced to take up one or more stains to make it more clearly visible. The stains used in light MICROSCOPY are coloured dyes and they heighten the differences between various tissue and cell components. They are usually organic compounds with a positive and negative ion. A stain in which the colour comes from the positive ion (organic cation) is called a basic stain, e.g. haemotoxylin. A stain in which the colour comes from the negative ion (organic anion) is called an acidic stain, e.g. eosin. If the colour comes from both components it is called a neutral stain, e.g. Leishman's stain. Biological material that takes up acidic stain is described as acidophilic while that taking up basic stain is basophilic. Similarly the third category is neutrophilic.

Vital stains are those that colour living components of cells or tissues without harming them. Non-vital stains are used to colour material that is dead. Staining may be permanent or temporary and in the latter case the colour fades and tends to cause damage to the material. The stains used in ELECTRON MICROSCOPY are electron dense and use heavy metals such as uranium and lead rather than coloured dyes.

stamen a male reproductive organ of a flower, comprising a thin stalk or filament that bears an expanded portion, the anther, at its apex. The anther consists of two fused lobes each containing two pollen sacs. The pollen sacs are microsporangia (*see* SPORANGIUM), which produce microspores that become the pollen grains. It is described as a microsporophyll (*see* SPOROPHYLL).

standing crop the complete amount of living material in a population within a unit area, called the biomass and expressed in terms of energy. Standing crop values may vary in DECIDUOUS plants between different times of year.

stapes one of the tiny bones of the mammalian MIDDLE EAR. It is known as the stirrup. It is the third EAR OSSICLE and it articulates with the INCUS (anvil) which in turn articulates with the MALLEUS (hammer). Vibrations of the eardrum (tympanic membrane) are conducted through the ear ossicles to the oval window which is a membrane beneath the stapes.

Staphylococcus a genus of bacteria that has a spherical shape and is non-mobile. It is often found clustered irregularly together like a bunch of grapes. They are saprophytes or PARASITES and some cause diseases in man and animals, e.g. infections such as mastitis, boils and abscesses and food poisoning.

starch a polysaccharide (*see* SACCHARIDES) found in

all green plants. It is built up of chains of GLUCOSE units arranged in two ways, as amylose (long unbranched chains) and amylopectin (long cross-linked chains). Potato and some cereal starches contain about 20-30 per cent amylose and 70-80 per cent amylopectin.

statocyst a sensory mechanoRECEPTOR present especially in invertebrates that functions in the maintenance of balance. It consists of a small sac or vesicle filled with fluid containing grains of material such as sand or calcium carbonate, the *statoliths* or *otoliths*. When these move they come into contact with specialized hairs or SETAE, which excite sensory cells. These transmit information about the position of the whole or part of the body, often enabling the animal to right itself if it has been turned over. The semicircular canals in the EARS of mammals act in a similar way and are the corresponding organs in vertebrates.

stearic acid found as glycerides in many animal and vegetable fats, it is a solid saturated FATTY ACID.

stele (*plural* **stelae**) a group of tissues present in plant stems and roots comprising the VASCULAR TISSUE (xylem and phloem) and the endodermis and pericycle if this is present. The structure of the stele varies between different types of plant and is one of the distinguishing features. Also, it tends to form a solid core in ROOTS but is present as a hollow cylinder in STEMS.

stem the part of a plant that is often found above ground growing upwards towards the light. The STELE is in the form of a hollow cylinder with vascular bundles arranged in a ring, and the stem bears the leaves, buds and reproductive organs. Stems may become woody as a result of secondary thickening (*see* GROWTH RING) and may bear spikes, thorns or hairs. Specialized stems may occur below ground, e.g. RHIZOMES, BULBS, CORMS, and TUBERS. The vascular tissue of a stem conducts water, mineral salts and nutrients from the roots to the leaves and supports the plant. The stem itself may also be involved in PHOTOSYNTHESIS.

stereochemistry the part of chemistry that covers the spatial arrangement of atoms within a molecule (*see* ISOMER).

stereoisomerism isomerism due to the different arrangement in space of atoms within a molecule, giving ISOMERs that are mirror images of each other.

stereotaxis the movement or reaction of an organism in response to contact with a solid body.

steric hindrance the phenomenon whereby the arrangement in space of atoms in reacting molecules hinders or slows a chemical reaction.

sterile (1) an organism that is unable to reproduce and in mammals this may be for a variety of reasons. (2) an absence of living organisms namely micro-organisms such as bacteria (*see* AUTOCLAVE).

sternum (*plural* **sterna**) (1) the breastbone of terrestrial vertebrates, which lies in a VENTRAL position in the thorax and to which the ends of most of the ribs are attached. It also articulates with the CLAVICLE (collar bone) of the PECTORAL GIRDLE and in birds it bears the KEEL. It is not present in fish. (2) The CUTICLE on the ventral side of each segment of an arthropod (*see* ARTHROPODA), which is often a thickened plate.

steroids a group of LIPIDS with a characteristic structure comprising four carbon rings fused together.

The group includes the sterols (e.g. CHOLESTEROL), the BILE acids, some HORMONES, and vitamin D. Synthetic steroids act like steroid hormones and include derivatives of the glucocorticoids used as anti-inflammatory agents in the treatment of rheumatoid arthritis; oral contraceptives, which are commonly mixtures of OESTROGEN and a derivative of PROGESTERONE (both female sex hormones); anabolic steroids, e.g. TESTOSTERONE, the male sex hormone, which is used to treat medical conditions such as osteoporosis and wasting. However, much publicity surrounds the use of the anabolic steroids by athletes to increase muscle bulk and body weight, in contravention of the rules of the sports-governing bodies.

sterols a subgroup of STEROIDS being steroid based alcohols consisting of saturated (*see* SATURATED COMPOUNDS) HYDROCARBONS with 17 carbon atoms in a ring system. They have a hydroxyl group at C3 and a side chain of eight to ten carbon atoms at C17. The most familiar animal sterol (zoosterol) is CHOLESTEROL.

stigma *see* **carpel**

stipe a stalk present in certain fungi, supporting the cap and forming the lower part of the fruiting body, e.g. mushroom. Also describes a stalk found in certain seaweeds, such as KELP, between the blade and the holdfast of the THALLUS.

stipule a structure arising from the base of the leaf stalk in many plants. It is often leaf-like and protects an axillary bud, and may also be photosynthetic as in the garden pea. It may be in the form of a spine, as in the false acacia (*Robinia*) or a scale, as in the lime tree.

stoichiometry an aspect of chemistry that deals with the proportion of elements (or chemical equivalents) making pure compounds.

stolon a long stalk that grows out horizontally from the parent plant and when it touches the soil the bud at the apex gives rise to a new plant. An example is the blackberry. *Layering* is a horticultural technique in which stolons are pinned down onto the soil surface to encourage the growth of new plants.

stoma (*plural* **stomata**) a pore found in the epidermis of plants occurring particularly on the leaves where stomata occur in large numbers (especially on the underside). Stomata are also common on young shoots. Each stoma is surrounded by a pair of specialized epidermal cells, known as guard cells, which have a crescent shape. The movements of these, which vary according to the water content (turgidity), alter the size of the pore aperture and control its opening and closing. The stomata are important in the process of TRANSPIRATION and are also the structures through which gaseous exchange takes place. The term stoma also describes both the pore and the guard cells.

stomach an expansion of the ALIMENTARY CANAL in vertebrates that lies between the OESOPHAGUS and SMALL INTESTINE. It has thick walls containing SMOOTH MUSCLE, which contracts to manipulate the food, and its exits are controlled anteriorly by the cardiac SPHINCTER and posteriorly by the pyloric sphincter. It is capable of significant expansion and contraction and mucosal cells in the stomach lining secrete gastric juice containing mucus, ENZYMES (pepsinogen) and hydrochloric acid. The acidic mass of partially digested food processed by the stomach is passed on to the small intestine where further digestion and absorption takes place.

stomium a part of the SPORANGIUM of a fern, where thin walled cells occur and which ruptures when ripe to release the spores.

stratopause the top of the stratosphere at about 50 kilometres.

stratosphere the layer of the atmosphere above the TROPOSPHERE, which stretches from 10 to 50 kilometres above the ground. It is a stable layer with the TROPOPAUSE at the base. The temperature increases from the lower part to the upper, where it is 0°C, and the higher temperatures are due to ozone-absorbing ultraviolet radiation. The inversion of temperatures creates the stability that tends to limit the vertical extent of cloud, producing the lateral extension of, for example, cumulonimbus cloud into the anvil head shape.

stratum corneum the outermost layer of the SKIN in mammals consisting of dead keratinized cells that are eventually sloughed off. It has a protective and waterproofing function.

Streptococcus a genus of bacteria that do not produce spores and have a spherical shape. They are commonly found in a variety of habitats, often forming long chains. They are saprophytes and PARASITES. They are found, for example, in the respiratory tract mucous membranes, intestines and on the skin of mammals, and in milk. Many are harmless but others are PATHOGENIC and cause infections such as scarlet fever, pneumonia and bacterial arthritis.

streptomycin an ANTIBIOTIC that inhibits the synthesis of proteins in PROCARYOTE cells (organisms belonging to the kingdom Monera). It inhibits the translation of the mRNA of the RIBOSOMES and is used in cell culture to help prevent the growth of micro-organisms that would cause contamination.

striated muscle a type of muscle found in vertebrates, also known as voluntary muscle as it is under conscious control. It is the type of muscle that operates the skeleton and consists of bundles of elongated fibres containing many nuclei and surrounded by a strong sheath of CONNECTIVE TISSUE, called the epimysium. A tendon at the end of the muscle attaches it to the bone. The *origin* of the muscle is the end that is attached to a non-moving bone whereas the *insertion* is the end attached to a moving bone. Each striated muscle fibre itself consists of smaller fibres or fibrils running longitudinally and called *myofibrils*. These have alternating light and dark bands called *sarcomeres* containing contractile proteins, ACTIN and MYOSIN. The sarcomeres confer the striated appearance and the contractile function of the muscle. Each fibre is surrounded by a membrane called the *sarcolemma*, consisting of a plasmalemma and basement membrane. A flexor muscle contracts and becomes shorter and fatter and moves one bone closer to another. This is also known as the *agonist*. An extensor or *antagonist* muscle operates in the opposite direction to move one bone away from another. While one muscle is contracting, the other is relaxed, becoming long and thin (*see also* MUSCLE).

stridulation the production of sound by some insects, which is typical of the ORTHOPTERA. Various parts of the body are involved depending upon the species, and usually one part is specialized, to be rubbed against another. The sound is used in courtship, warning and territorial behaviour.

stroma (*plural* **stromata**) any tissue that functions as a framework in plant cells (*see* CALVIN CYCLE).

stromatolite laminated calcareous structures produced by calcareous algae (cyanophyte or cyanobacteria) that trap and bind fine sediment, to produce in most cases a structure with some vertical dimension, e.g. domes, columns. Stromatolites are found today in tropical carbonate environments in shallow water but in the past their distribution was more widespread. Current forms have laminae millimetres thick and an overall length of centimetres while Precambrian varieties are known several metres thick.

strophism the twisting, by progressive growth, of a stalk in response to a stimulus from a particular direction, e.g. light.

structural formula a formula providing information on the ATOMS present in a MOLECULE and the way that they are bound together, i.e. an indication of the structure.

strychnine a crystalline ALKALOID with a very strong bitter taste and a very dangerous action upon the nervous system.

style *see* **carpel**

subarachnoid space a space between two of the membranes (MENINGES) surrounding the brain and spinal cord, the pia mater and the arachnoid. It is filled with CEREBROSPINAL FLUID.

subclavian artery one of a pair of arteries in vertebrates that run beneath the CLAVICLE (collar bone), the one on the right arising from the inominate artery and the one on the left from the AORTA. They supply blood to the arm.

subcutaneous tissue lying immediately beneath the DERMIS of the SKIN in vertebrates and consisting of loose connective tissue, adipose tissue, blood vessels, muscles and nerves. A sheet of STRIATED MUSCLE may be present, which moves the scales or the skin and also small muscles to erect the hair (erector pili muscles). The fat that is stored provides insulation and a food store that is important in some mammals, such as those living in very cold conditions and those that hibernate.

substrate in biology (1) the surface upon which an organism lives and from which it may derive its food. (2) a substance/reactant in a reaction that is catalysed by an ENZYME.

succulent one of a number of plants that stores water in fleshy leaves or stems, e.g. cacti. Succulents are adapted to grow in very arid conditions or those in which fresh water is not easily available. They are often modified to reduce water loss, in addition to their capacity for storage.

succus entericus the digestive juice secreted by the crypts of Lieberkuhn glands that occur between the VILLI of the SMALL INTESTINE. It contains enzymes— amylases, LIPASES and proteases—that complete the digestive processes. About two to three litres per day is secreted by the human small intestine.

sucrose a disaccharide CARBOHYDRATE ($C_{12}H_{22}O_{11}$) occurring in beet, sugar cane and other plants (*see* SACCHARIDES).

sugar a crystalline monosaccharide or oligosaccharide (a small number, usually two to ten, monosaccharides linked together, with the loss of water), soluble in water. The common name for SUCROSE.

sulphonamides *also known as* **sulpha drugs** a group of drugs with antibacterial action. They are amides of sulphonic acids and have the group $-SO_2NH_2$. They are used medically to combat infections, especially of the urinary tract and gut. They operate by preventing the multiplication of bacteria.

summation the process by which pre-synaptic impulses are added together until an impulse is generated in the post-synaptic membrane. It occurs when one or few pre-synaptic impulses are not sufficient to initiate a response whereas a train of them can. It is called *temporal summation* when the impulses are added together at the same SYNAPSE and *spatial summation* when they arrive at different synapses made by a single cell. It is one of the mechanisms for exerting fine control over the responses mediated by the nervous system (*see also* NEUROMUSCULAR JUNCTION).

surface tension a tension created by forces of attraction between molecules in a liquid, resulting in an apparent elastic membrane over the surface of the liquid.

surfactant (*also called* **surface-active agent**) a compound that reduces the SURFACE TENSION of its solvent, e.g. a detergent in water.

suture a line marking the fusion of two adjoining structures in a plant or animal body. Examples are the suture lines of the skull in vertebrates and in plants, the seam along a pod such as that of a pea or bean.

sweat a salty secretion of sodium chloride, sodium lactate and urea produced by the SWEAT GLANDS and secreted onto the skin surface. Sweat is one of the cooling mechanisms of mammals as the liquid is evaporated by the body heat, thus cooling the body.

sweat gland the glands present in the EPIDERMIS of mammalian SKIN that project into the DERMIS and are under the control of the SYMPATHETIC NERVOUS SYSTEM. They are responsible for the secretion of SWEAT and their distribution in the skin varies according to the species of mammal. There are also two types: apocrine and eccrine sweat glands.

swim bladder present in BONY FISH (Osteichthyes), this is an air-filled sac located above the alimentary canal. It regulates the buoyancy of the fish, air entering either via blood capillaries or through a duct that opens into the stomach or oesophagus. It has developed its hydrostatic function through evolutionary specialization, being the organ that corresponds to the lungs in other vertebrates, and it is still used for gaseous exchange in lungfish.

symbiosis a relationship between organisms, usually two different species, that has beneficial consequences for at least one of the organisms. There are various forms of symbiosis, including commensalism, where one party benefits but the other remains unharmed, and parasitism, where one party greatly benefits (the PARASITE) at the other party's expense. Symbiosis can also solely refer to mutualism, where both parties benefit and neither is harmed.

symbol a letter or letters that represent an element or an atom of an element (*see* element table in APPENDIX 2 for list of symbols).

symmetry the property of a geometrical figure whose points have corresponding points reflected in a given line (axis of symmetry), point (centre of symmetry, e.g. a circle) or plane (reflection).

sympathetic nervous system along with the PARA-SYMPATHETIC NERVOUS SYSTEM, this makes up the AU-TONOMIC NERVOUS SYSTEM. It acts in opposition to the parasympathetic nervous system and noradrenaline and adrenaline are the main NEUROTRANSMIT-TERS released by its nerve endings. Some of its functions include raising the heart beat rate, inhibiting the secretion of saliva and constricting blood vessels.

synapse (*see also* NEURON) the junction between two nerve cells where a minute gap (of the order of 15 NANOMETRES) occurs. The nerve impulse is carried across the gap by a neurotransmitter substance (*see* ACETYLCHOLINE). The chemical diffuses across the gap connecting the axon of one nerve cell to the dendrites of the next. An individual neuron commonly has several thousand such junctions with around 1000 other neurons.

synergism is when the combined effect of two substances (e.g. drugs) is greater than expected from their individual actions added together.

synovial membrane *see* **joint**.

synthesis the formation of a compound from its constituent elements or simple compounds.

syrinx a complex sound-producing organ found in birds and located at the lower end of the trachea where the latter divides to form the BRONCHI. Sound is produced by means of vibrating membranes of connective tissue—the vocal chords of the bird (*see also* LARYNX).

systolic blood pressure the pressure generated by the left VENTRICLE of the HEART at the peak of its contraction. Since the left ventricle has to pump blood to all parts of the body, it generates a higher pressure than the right ventricle, which pumps blood only to the lungs. In normal people, the systolic blood pressure is 120 mm of mercury (120 mm Hg), and when the ventricle relaxes, pressure is still maintained in the blood vessels. This resting pressure is called the diastolic pressure and is approximately 80 mm Hg. This is the familiar blood pressure measurement and is represented as 120/80. This fluctuation in pressure is responsible for the pulse, which also represents the heartbeat.

T

tannin one of a group of compounds common in leaves, tree bark and unripe fruits. These astringent organic chemicals include phenols and their function is unclear, they may play a role in discouraging grazing animals. Some tannins are used commercially in leather production.

tapetum a layer of GUANINE crystals in the choroid (a thin, pigmented layer behind the RETINA) of the eye of nocturnal vertebrates. It improves vision by reflecting light back onto the retina (and makes eyes shine in the dark).

tapeworms *see* **Cestoda**.

tarsal bones bones that form the ankle (tarsus) in terrestrial vertebrates. In man there are seven bones and one forms the heel. They articulate with the TIBIA and FIBULA and the METATARSALS.

taste bud a taste sense organ found in most vertebrates and comprising a group of cells usually located on the tongue. Four tastes can be differentiated—salt, sweet, sour and bitter and on the human tongue there is a particular pattern of these taste buds. In aquatic animals the buds are often widely scattered and fish have buds over the whole body to gather information about the water.

taxis (*plural* **taxes**) the movement of a cell or organism in response to a stimulus in the environment. This stimulus may be temperature (thermotaxis), light (phototaxis), gravity (geotaxis) or chemical (CHEMOTAXIS).

taxonomy the study, identification and organization of organisms into a hierarchy of diversity according to their similarities and differences. Taxonomy is concerned with the classification of all organisms, whether plant or animal, dead or alive, e.g. fossils are also important. Modern taxonomy provides a convenient method of identification and classification of organisms, which expresses the evolutionary relationships to one another. Various aspects and characteristics of organisms are used: morphology and anatomy (*classical taxonomy*), protein structure (*biochemical*), chromosomal (*cytotaxonomy*), and *numerical* taxonomy assesses, mathematically, the division into groups.

T-cell a type of white blood cell (LYMPHOCYTE) that differentiates in a gland called the thymus, situated in the thorax. There are a whole variety of T-cells involved in the recognition of a specific foreign body (ANTIGEN), and they are particularly important in combating viral infections and destroying bacteria that have penetrated the cells of the body.

telophase the last stage of MEIOSIS or MITOSIS in EUCARYOTIC cells. During telophase, a nuclear membrane forms round each of the two sets of CHROMOSOMES that have formed separate groups at the spindle poles. The chromosomes decondense, the nucleoli reappear, and the cell eventually splits to form two daughter cells.

temperature degree of heat or cold against a standard scale.

tendon a cord or sheet of connective tissue attaching muscle to bone. It comprises inelastic COLLAGEN fibres and it makes sure that the relevant movement accompanies a muscle contraction.

tendril a modified stem or leaf that becomes a thin filamentous structure and that is used as a support by climbing plants. The tendrils respond to and wrap around solid objects, the twisting being accomplished by the cells in contact with the object losing water and curving. It is notable in pea plants.

tensor a muscle that stretches a part of the body without altering the relative position of that part.

teratogen anything that may result in malformation of an embryo. Included are IONIZING RADIATIONS, drugs, toxic chemicals, viral infections, etc.

terpenes colourless, liquid hydrocarbons occurring in many fragrant natural oils of plants. The general formula is (C_5H_8)n where C_5H_8 is the basic isoprene unit. This leads to their classification: monoterpenes are $C_{10}H_{16}$; sesquiterpenes $C_{15}H_{24}$; diterpenes $C_{20}H_{32}$, and so on.

terpenoids a group of plant SECONDARY METABOLITES based on isoprene units (C_5; *see also* TERPENES). The group includes many essential oils, the CAROTENOIDS, rubber, and the GIBBERELLINS, which are plant growth substances promoting shoot elongation in some plants and promoting seed germination.

territory a specific area that animals, whether singly or in groups, defend to exclude other members of their species. A territory may be used for feeding, mating and breeding or all these activities and the size varies enormously from the small nesting territories of sea birds to the large areas used by red squirrels. Many mammals use scent markings to delineate their territories while birds sing to indicate the limits of their territory. Territory differs from home range, the latter being the area in which the animal roams and which is not defended.

Tertiary the oldest period of the Cenozoic era, from 65 to 24.6 million years ago and comprising the PALAEOCENE, EOCENE, OLIGOCENE, MIOCENE and PLIOCENE. It was notable for the rise of modern mammals and the development of grasses and flowering plants.

testa (*plural* **testae**) the protective covering of a seed. It is formed from the INTEGUMENTS of the OVULE after fertilization.

testis (*plural* **testes**) the male reproductive organ that produces sperm (SPERMATOZOA) and also steroid hormones (ANDROGENS). In mammals, the testes are contained in the sac-like SCROTUM outside the body. Within the testis are numerous seminiferous tubules, in which sperm develop, which pair with the VAS DEFERENS.

testosterone a male sex hormone that promotes the development of male characteristics.

tetanus a sustained muscle contraction (*see* ACTION POTENTIAL). The word is also used as the name of a disease caused by a toxin from the bacterium *Clostridium tetani*. It causes increasing muscle spasms that make it difficult to open the mouth (hence the term "lock-jaw"), and it may be so severe that the sufferer dies of asphyxia or exhaustion.

tetrapod a four-limbed vertebrate. Included are mammals, birds, reptiles and amphibians. The pentadactyl limb is the basis for the limbs of all tetrapods.

thalamus a part of the FOREBRAIN in vertebrates that

receives sensory information from all the senses and relays it to the CEREBRUM. It also receives signals from the cerebrum and other parts of the brain that control emotion.

thallus (*plural* **thalli**) a simple plant body without subdivision into roots, stems and leaves. It is found in ALGAe, (e.g. seaweed), LICHENS and BRYOPHYTES.

thermocline in lakes and oceans, the surface of greatest drop of temperature with depth. In summer months two layers with significantly different temperatures are formed—an upper warm layer of circulating water and a deeper cold, relatively undisturbed region (the epilimnion and hypolimnion respectively). The thermocline occurs over a layer between the two.

thermodynamics the study of laws affecting processes that involve heat changes and energy transfer. There are essentially three laws of thermodynamics. The first (the law of conservation of energy) states that within a system, energy can neither be created nor destroyed. The second law says that the ENTROPY of a closed system increases with time. The result of the third law (or Nernst heat theorem) is that absolute zero can never be attained.

thermography the medical scanning technique whereby the infrared radiation or radiant heat emitted by the skin is photographed, using special film, to create images. An increase in heat emission signifies an increase in blood supply, which may be indicative of a CANCER. The technique is used to detect cancers, especially of the breast.

thermoluminescence a phenomenon whereby a material emits light upon heating because of ELECTRONs being freed from defects in crystals. The defects are generally due to ionizing radiations, and the principle is applied in dating archaeological remains, especially ceramics, on the assumption that the number of trapped electrons, caused by exposure to radiations, is a function of time. Although this is not absolutely correct, an estimate of the age of a piece of pottery can be obtained by heating and comparing the thermoluminescence with that of an item of known age.

thermometer an instrument used to measure temperature. Any property of a substance, providing it varies reliably with temperature, may form the basis of a thermometer. This includes expansion of liquids or gases, or changes in electrical RESISTANCE.

thiamine (*see also* VITAMIN B COMPLEX) the first vitamin (B_1) to be discovered. A deficiency causes BERI-BERI. Thiamine can be obtained through eating pork, whole grains and legumes.

thin layer chromatography CHROMATOGRAPHY that occurs essentially in two dimensions with the separation of small quantities of mixture achieved by movement of a solvent across a flat surface on sheets of absorbent paper or special materials (e.g. silica gel supported on glass plates). The components of the sample move at different rates across the "plate" because of differences in solubility, size, charge, etc, and after the process, the components can be examined *in situ* or removed for further analysis. The technique has its disadvantages but it is used widely for checks on purity or to characterize complex materials.

thoracic duct the main LYMPH collecting vessel in mammals. The duct runs up the thorax close to the vertebral column and collects lymph from the trunk and hindlimbs. The lymph is drained into the superior vena cava.

thoracic vertebrae the vertebrae of the upper part of the back, between the cervical and LUMBAR VERTEBRAE. The thoracic vertebrae, of which there are twelve in man, articulate with the ribs.

thorax the body cavity that, in terrestrial vertebrates, contains the heart and lungs within the rib cage and is separated from the abdominal cavity by the diaphragm. In insects (*see* INSECTA) it is the body between the head and abdomen, which, in adults, often forms three segments. In some groups, e.g. the crustaceans, the thorax and head fuse forming the cephalothorax.

Thornthwaite climate classification (*see also* KÖPPEN CLASSIFICATION) a classification of climates devised in the early 1930s by C. W. Thornthwaite and based upon characteristic vegetations and the associated precipitation. Two parameters were developed: P/E and T/E. P/E represents the total monthly precipitation divided by the total monthly evaporation and T/E is similar save that it refers to the temperature. Based upon the precipitation figures, five humidity provinces were created:

province	vegetation	P/E
A	rain forest	>127
B	forest	64–127
C	grassland	32–63
D	steppe	16–31
E	desert	< 16

Each province can be further subdivided by the rainfall pattern, e.g. abundant in all seasons, deficient in winter. The T/E figures provide an additional classification into provinces:

	province	*T/E*
A'	tropical	128
B'	mesothermal	64–127
C'	microthermal	32–63
D'	taiga	16–31
E'	tundra	1-15
F'	frost	0

With all this information, Thornthwaite produced a climatic world map of 32 climate types. Each of these areas can be associated with certain soils, geomorphological processes, etc, related to and depending upon the climate.

threshold the critical level that a stimulus must exceed in order to produce a response in a nerve or muscle (*see also* ACTION POTENTIAL).

thrombin an enzyme involved in the process of BLOOD CLOTTING. It catalyses the production of FIBRIN from FIBRINOGEN.

thymine ($C_5H_6N_2O_2$) a nitrogenous base component of DNA that has the structure of a PYRIMIDINE. Thymine always base-pairs with ADENINE in a DNA molecule, but in RNA molecules it is replaced with URACIL.

thymus an organ found only in vertebrates, situated near to the heart in the upper chest. It is instrumental in the development of lymphoid tissue, white cells that produce antibodies and thus the IMMUNE SYSTEM. The thymus gland shrinks in size at puberty (when the immune system is well established) and by adulthood has almost disappeared.

thyroid a gland with two lobes, found at the base of the neck in humans and other mammals. This endocrine gland produces two hormones that contain iodine, thyroxine and triiodothyronine (*see* THY-

ROID HORMONES). The thyroid is vital as it regulates metabolism and also influences physical development. Deficiency produces marked effects, including cretinism (retarded skeletal growth and poor mental development) and goitre (an enlarged thyroid due to iodine deficiency). Thyroid growth and activity is controlled by the anterior pituitary gland through a hormone—thyrotrophin (or thyroid-stimulating hormone, TSH). In amphibians, the thyroid controls the METAMORPHOSIS of a tadpole into a frog.

thyroid hormones thyroxine and triiodothyronine are two hormones secreted by the THYROID, the former in greater quantities, although the latter is more potent. If hormone levels or blood temperature are too low the hypothalamus releases thyrotropin releasing factor into the anterior pituitary to prompt release of TSH (*see* THYROID), which releases the hormones from the thyroid. The hormones increase the metabolic rate and raise body temperature. The thyroid also releases calcitonin, which lowers calcium and phosphate concentrations in the blood.

tibia the shinbone,, which with the FIBULA, forms the lower hindlimb of tetrapod vertebrates. It articulates at the base with the FEMUR and at the ankle with the TARSUS.

tissue a group of cells with a similar function, which aggregate to form an organ.

tissue culture the culture or growth, outside the body, of TISSUES of living organisms. The cells are placed in an artificial medium containing nutrients and other factors such as temperature and pH are controlled, while waste products are removed. The technique has proven valuable in studying cell growth and it has been applied to the propagation of plants.

toadstool the general name for the fruiting bodies of fungi but not mushrooms. Some are poisonous.

tocopherol *see* **vitamin E**.

tomography a scanning technique that uses X-rays for photographing particular "slices" of the body. A special scanning machine rotates around the horizontal patient, taking measurements every few degrees over 180°. The scanner's own computer builds up a three-dimensional image that can then be used for diagnosis. Such a technique has the dual benefit of providing more detail than a conventional X-ray and yet delivers only one fifth of the dose.

tone muscle tension that is required to maintain posture. The muscles used for posture respond in a different mode to that for ordinary contractions. The fibres contract more slowly, and only a proportion of the fibres contract, sequentially, so that noncontracting fibres can recover for further contractions.

tonne a metric ton (1000 kilograms).

tonsils lymphoid tissue, at the back of the mouth and throat in higher vertebrates, that is involved with LYMPHOCYTE production. Humans have *palatine* tonsils at the back of the mouth, *lingual* tonsils beneath the tongue and *pharyngeal* tonsils (or adenoids).

tooth (*plural* **teeth**) a hard structure used by animals for biting and chewing food, for attacking, for grooming, and for other functions. In higher vertebrates, teeth are concentrated in the jaws, but fish and amphibians have teeth all over the palate. Teeth evolved from scales of cartilaginous fish and comprise a central PULP CAVITY with nerve and blood supply, surrounded by DENTINE and ENAMEL on the exposed surface. The *root* is embedded in the jaw bone. Mammals have four types of teeth (MOLAR, PREMOLAR, CANINE and INCISOR). Numbers of teeth and their arrangement differ between animals (*see* DENTITION).

totipotency the capacity of a cell to generate all the characteristics of the adult organism. Totipotent cells have full genetic potential, unlike most adult cells, which have lost this during the process of differentiation when they form cells with specialized functions.

toxin any poison produced by a living organism, whether from a BACTERIUM, plant or animal. An *endotoxin* is one released by dying bacteria, whereas an *exotoxin* is secreted into the surrounding medium. The toxin acts as an ANTIGEN in the body.

trace element elements required in minute amounts for healthy growth (*see* ESSENTIAL ELEMENT).

trace fossil a sedimentary structure caused in some way by the presence or activity of an organism. The study of trace fossils is called ichnology. They are most common at the junction of different lithologies, for example shale and sandstone. The fossils may occur as ridges, tubes, burrows, etc.

trachea (*plural* **tracheae**) the windpipe in (air-breathing) vertebrates. Its shape is maintained by rings of cartilage and air passes from the throat to the BRONCHI. In insects tracheae are fine branching tubes, which subdivide into finer tracheoles and through which air is pumped. The external openings are called SPIRACLES.

trait a specific phenotypic character (*see* PHENOTYPE). Thus character, such as flower colour, is a heritable feature but each variant for the character (yellow or white flowers, etc) is a trait.

transamination a biochemical reaction involving the transfer of an AMINO group from an AMINO ACID to a keto acid (such as pyruvate) forming a new amino acid and keto acid.

transcription the formation of an RNA molecule from one strand of a DNA molecule. Transcription involves many processes, starting with the unwinding of the double-stranded DNA helix, along which an enzyme, called RNA polymerase, travels and catalyses the formation of the RNA molecule by pairing NUCLEOTIDEs with the corresponding sequence of the DNA strand. As the RNA molecule leaves, the DNA reforms its double-stranded helix.

transect in ecological studies, a straight line across an area, along which measurements are taken and data recorded.

transferase an ENZYME that catalyses the transfer of chemical groups between molecules, e.g. acyl transferase.

transfer RNA (tRNA) one of the three major classes of RNA that functions as the carrier of AMINO ACIDS to RIBOSOMES, where the POLYPEPTIDE chains of PROTEINS are formed. Every tRNA molecule has a structure that will accept only the specific attachment of one amino acid.

transgenic a term for an organism that contains genes from another species, created by genetic engineering. There have been a number of transgenic animals developed in agriculture, for example extra growth hormone can be produced to improve meat yield. Also, plants can be made resistant to certain diseases using transgenic methods.

translation the synthesis of PROTEINS in a RIBOSOME that has MESSENGER RNA (mRNA) attached to it. As the mRNA molecule moves through part of the ribosome, a TRANSFER RNA molecule carrying the appropriate AMINO ACID will enter a site on the ribosome and will be released after it has contributed a new amino acid to the growing chain.

translocation the two processes of movements of compounds in plants. One involves the absorption of mineral ions from solution in the soil and their transport via the XYLEM to other parts of the plant. The second process is the movement in the PHLOEM, of compounds synthesized by the leaves up and down the plant to the growing points in particular.

transpiration the loss of water vapour from pores (STOMAta) in the leaves of plants. Transpiration can sometimes account for the loss of over one sixth of the water that has been taken up by the plant roots. The transpiration rate is affected by many environmental factors—temperature, light and carbon dioxide (CO_2) levels, air currents, humidity, and the water supply from the plant roots. The greatest transpiration rate will occur if a plant is photosynthesizing (*see* PHOTOSYNTHESIS) in warm, dry and windy conditions.

transposon segments of DNA that can move to different locations in a cell's GENOME—a mobile genetic element. The simplest case is in bacteria, where a transposon moves from one CHROMOSOME locus to another or from a PLASMID to a chromosome. Complex transposons may combine several genes for resistance to antibiotics into one plasmid by moving the genes to that position.

Trematoda a class of flatworms within the phylum

PLATYHELMINTHES. The trematodes are parasitic (*see* PARASITE) and comprise flukes such as the blood fluke *Schistosoma* in man, and the liver fluke of sheep, *Fasciola*. Most flukes have suckers or hooks for attachment to the host and a protective cuticle to the body. The life cycles often involve intermediate hosts where the larvae develop before the final host is infected.

Triassic the first period of the Mesozoic era, lasting from 248 to 213 million years ago. After a mass extinction in the late Palaeozoic, new flora and fauna appeared. Molluscs, ammonites and bivalves were all abundant and reptiles (DINOSAURS, turtles and ichthyosaurs) the dominant vertebrates. There were also some GYMNOSPERMS.

tricuspid valve a valve between the atrium and ventricle on the right side of the mammalian and bird hearts. It consists of three flaps that close to prevent blood flowing back into the atrium when the ventricle contracts.

trilobites a class of primitive arthropods that became extinct in the Permian. There are almost 4000 identifiable species and after their appearance in the Cambrian (*see* APPENDIX 5) they became common in Palaeozoic seas. The body was divided into three parts (essentially into the head, thorax and tail) and also into three lobes running along the length of the body. The trilobite eye resembles the compound eye of living arthropods. On average, trilobites were up to 10cm long although exceptional specimens up to one metre did exist.

trisomy the abnormal condition in which an organism has three CHROMOSOMES rather than the normal pair for one type of chromosome. Trisomy can occur in humans and results in offspring with abnormal characteristics and shorter-than-average lifespans. One common example of trisomy is Down's syndrome, caused by the presence of three instead of two chromosomes of the number 21 type (all the other chromosomes are in normal pairs).

tRNA *abbreviation for* TRANSFER RNA.

trophic relating to nutrition.

trophic level (*see also* FOOD CHAIN) a concept in ecology that places plants and animals on different levels in the food chain. At the base are the primary producers (plants) followed by herbivores; primary, secondary and tertiary carnivores; and decomposers. Many animals do not feed at only one trophic level.

tropical rain forest one of the tropical forest BIOMES, found in three major regions of the world. The largest is in the Amazon basin, with others in Africa (the Zaire basin) and Sri Lanka, India, Thailand, and the Philippines. There is very high rainfall and an enormous diversity of species with up to 300 tree species, some reaching 61 metres (200 feet). Because of the abundance of trees, competition for light is intense, and the leaves form a closed canopy, restricting access of light to the forest floor. The animals present usually live in the tree tops and include monkeys, insects, birds, snakes and bats. The high temperatures suit numerous reptiles and amphibians.

Organic debris tends not to build up because of rapid breakdown by decomposers, the released nutrients being quickly taken up by roots, or washed away by rain. The soil is thus relatively infertile. This infertility is enhanced by the actions of human beings in destroying large swathes of forest. After clearing, leaching of remaining nutrients leads to soil removal or hardening rendering cultivation impossible. This may ultimately lead to the production of desert conditions, with ever-increasing detrimental effects upon the global climate.

tropism growth of a plant organ in a particular direction due to an external stimulus, e.g. touch, light.

tropopause the top of the TROPOSPHERE. The boundary between the troposphere and the stratosphere. The altitude of the tropopause differs over short periods, from 9–12 kilometres (5.5–7.5 miles) over the poles to about 17 kilometres (10.5 miles) over the equator.

troposphere the Earth's atmosphere between the surface and the TROPOPAUSE. This layer contains most of the water vapour in the atmosphere and most of the aerosols in suspension. The temperature decreases with height at approximately 6.5°C per kilometre and it is in this layer that most weather features occur.

trypanosomes flagellated protozoans within the phylum Zoomastigophora of the kingdom PROTISTA. Species of *Trypanosoma* cause African sleeping sickness, a disease spread by the bite of the tsetse fly.

trypsin an enzyme that is active in the digestion of proteins in the small intestine. It is secreted initially in the inactive form, trypsinogen, into the DUODENUM, from the PANCREAS. The trypsinogen is changed into trypsin by another enzyme, enterokinase.

tuber an underground swelling of a root or stem in some plants, e.g. the potato (a stem tuber). The following season new plants develop from the buds (or eyes). Root tubers (e.g. the dahlia) are for storage of food and may also produce new plants. The tuber is a means of surviving winter or a dry season in addition to its propagative qualities.

tundra a zone within the periglacial areas of Earth (i.e. originally those areas that bordered Pleistocene ice sheets, but now the term is used to signify environments subjected to freezing and thawing of the land) comprising treeless plains that, because of PERMAFROST conditions, have soils that are water-logged. The dominant plants are grasses, rushes, and sedges, with some herbs and dwarf woody plants.

Turbellaria a class of free-living worms within the phylum PLATYHELMINTHES. These flatworms are not parasitic, and are mainly marine, although the *planarian* species are common in streams and ponds. They are simple animals with no organs for respiration or circulation and very primitive excretory and digestive systems. Because they are small and flat, most cells are close to the water in which they live, facilitating exchange of gas and nitrogenous waste. They are carnivorous, and after covering their prey (or carrion) with digestive juice, suck up the food into a ramified gastrovascular cavity in which the digestive process continues. Planarians move using CILIA on their underside or by undulating the body. The head has eyespots and a nervous system more complex than that of CNIDARIANs (they can "learn" in a fundamental way). Both sexual and asexual reproduction occurs, the latter through regeneration.

turgor the term describing a plant cell when the VACUOLE increases its volume and the protoplasm is pushed up to the cell wall. In this *turgid* state the osmotic pressure (*see* OSMOSIS) pushing in water becomes balanced by the force exerted by the cell wall, and this helps a plant stay rigid and not wilt.

Turner's syndrome a genetic disorder that renders women sterile due to lack of ovaries. Secondary sexual characteristics (e.g. pubic hair) are missing and there is no menstrual cycle although the external genitalia are present. It is caused by the absence of one X-chromosome.

tympanum *or* **tympanic membrane** the membrane between the outer and MIDDLE EAR, which vibrates in sound waves. It transmits the vibrations to the *ear ossicles* (and thence onto the COCHLEA). Some animals, e.g. amphibians, have no external ear and the tympanum is at the surface of the skin.

type specimen *see* **holotype**.

U

ulna the bone, that with the smaller RADIUS, makes up the forelimb of tetrapod vertebrates. It articulates with the carpels at the wrist and at the elbow with the HUMERUS. The protrusion of the ulna above the humerus forms the olecranon process, better known as the "funny bone".

ultracentrifuge a machine that generates high centrifugal forces as its rotor is capable of spinning at speeds of up to 50,000 revolutions per minute. The ultracentrifuge is most commonly used during the separation of the various ORGANELLEs within cells. The larger and more dense organelles will form a deposit in the centrifuge tube more readily than the smaller, less dense ones. Thus the largest organelle of any normal cell, the NUCLEUS, will be deposited at the bottom of a centrifuge tube when the ultracentrifuge spins at a force of 600g for 10 minutes, whereas the smaller MITOCHONDRION needs a higher speed of 15,000g for 5 minutes to be deposited.

ultrasonic a term used to describe sound waves that are inaudible to humans as they have a frequency above 20kHz. Although the human ear is incapable of detecting such a high frequency, some animals, such as dogs and bats, can detect ultrasonic waves (*also known as* **ultrasound**). Ultrasound is used widely in industry, medicine and research. For example, it is used to detect faults or cracks in underground pipes and to destroy kidney stones and gallstones. The most recent development in ultrasonics is their use in chemical processes to trigger reactions involved in the production of food, plastics and antibiotics. Ultrasonics make certain chemical processes safer and cheaper as they eliminate the need for high temperatures and expensive catalysts.

ultraviolet radiation a form of radiation that occurs beyond the violet end of the visible light spectrum of electromagnetic waves. Ultraviolet rays have a frequency ranging from 10^{15}Hz to 10^{18}Hz, with a wavelength ranging from 10^{-7}m to 10^{-10}m. They are part of natural sunlight and are also emitted by white-hot objects (as opposed to red-hot objects, which emit infrared radiation). As well as affecting photographic film and causing certain minerals to fluoresce, ultraviolet radiation will rapidly destroy bacteria. Although ultraviolet rays in sunlight will convert steroids in human skin to vitamin D (essential for healthy bone growth), an excess can cause irreversible damage to the skin and eyes and can damage the structure of the DNA in cells by producing THYMINE-thymine dimers. Fortunately, a great deal of the ultraviolet radiation from the sun is prevented from reaching the earth as the OZONE LAYER in the upper atmosphere acts as a UV filter.

umbilical cord in placental mammals, the cord connecting the embryo to the placenta of the mother. It contains blood vessels (umbilical artery and veins) and is surrounded by amniotic fluid. At birth it is severed and shrivels to leave a scar, the navel.

ungulate a general term for hoofed mammals. There are two such orders, the Artiodactyla (cattle, camels, hippopotamus, pigs), and the Perissodactyla, which includes the horse.

unicellular single-celled, whether referring to an organ, organism or a cell. Organisms include the protozoans, bacteria, some algae, which are placed in a separate kingdom, and the PROTISTA.

unified scale (*see also* RELATIVE ATOMIC MASS) the scale that lists atomic and molecular weights using the ^{12}C isotope as the basis for the scale and taking the mass of the isotope as exactly twelve. This means the atomic mass unit is 1.660×10^{-27}kg.

universal indicator a mixture of certain substances, which will change colour to reflect the changing pH of a SOLUTION. Universal indicator is available in the form of a solution or paper strip and is used as an approximate measure of the pH of a solution by using the following chart as a guide:

Colour of indicator	red	orange	yellow	green	blue	purple
pH	1 2 3	4 5	6 7 8 9	10 11	12 13	14

unsaturated a chemical term used to indicate that a compound or solution is capable of undergoing a chemical reaction because of specific physical properties of the compound or solution. In the case of unsaturated organic compounds, the carbon atoms are unsaturated as they form double or triple bonds and are thus capable of undergoing addition reactions. If a SOLUTION is described as unsaturated, then it contains a lower concentration of SOLUTE dissolved in a definite amount of the SOLVENT than the concentration of solute needed to establish the EQUILIBRIUM found in a SATURATED solution.

uracil ($C_4H_4N_2O_2$) a nitrogenous base component of the NUCLEIC ACID, RNA, that has the structure of a PYRIMIDINE. During TRANSCRIPTION, uracil will always form a BASE PAIR with ADENINE of the DNA template, and during TRANSLATION, uracil will always base pair with adenine of the MESSENGER RNA (mRNA) molecule.

urea an organic molecule, $CO(NH_2)_2$, that is a metabolic byproduct of the chemical breakdown of PROTEIN in mammals. In humans, 20-30 grams of urea are excreted daily in the urine, and although urea is not poisonous in itself, an excess of it in the blood implies a defective kidney, which will cause an excess of other, possibly poisonous, waste products.

urea cycle (or ornithine cycle) a cyclical series of biochemical reactions that excretes nitrogen as UREA in mammals and most adult amphibians. Ammonia is the primary nitrogenous waste product but its toxicity dictates that for many animals it is converted to urea (or URIC ACID in birds). The conversion to urea occurs in the liver and it is then excreted in solution as URINE.

ureter in vertebrates, the duct carrying URINE out of the body from the KIDNEY to the BLADDER.

urethra in mammals, the duct carrying URINE from the KIDNEY to the BLADDER.

uric acid the main nitrogenous waste product in animals such as snakes, lizards, birds and insects etc. Animals employing this system are called URICOTELIC. It is a useful adaptation when conservation of water is necessary. Uric acid is insoluble in water and is excreted as a precipitate after most of the water has been reabsorbed.

uricotelic *see* **uric acid**.

uridine a molecule consisting of the nitrogenous base URACIL and the ribose sugar that is a basic unit of RNA structure when a phosphate group (H_3PO_4) is added to form a NUCLEOTIDE.

urine the fluid excreted by some animals to remove nitrogenous metabolic waste products. It is produced by the kidneys and excreted via the BLADDER and URETHRA (*see also* UREA CYCLE). In addition to water and the nitrogenous compound, urine may contain amino acids, purine and some inorganic ions. In contrast to aquatic animals that produce large volumes, terrestrial animals produce less to conserve water. Humans excrete about 1 to $1^1/_2$ litres per day.

uterus (or womb) in female mammals, the muscular organ between the BLADDER and RECTUM in which an embryo develops.It connects to the FALLOPIAN TUBES and VAGINA and the lining shows periodic changes (*see* MENSTRUAL CYCLE). Contractions of the muscular outer wall push the young out through the vagina at birth (*see* PARTURITION).

V

vaccine a modified preparation of a VIRUS or BACTERIA that is no longer dangerous but is capable of stimulating an immune response and thus confers immunity against infection with the actual disease. Vaccines can be administered orally or by a hypodermic syringe and are not effective immediately as it takes time for the recipient's IMMUNE SYSTEM to develop a memory for the modified virus or bacterium by producing specific ANTIBODIES.

vacuole a space in the cytoplasm of a cell that is filled with water, air, food particles, etc. In plants, the cells usually have one large vacuole that may contain sugar solutions, salts and some ions. Animal cells tend to have several small vacuoles.

vagina the passage leading from the uterus to the outside and in which sperm are deposited during copulation. The vagina is lined with stratified epithelium and in some mammals can be sealed when the animal is not sexually active.

vagus nerve an important part of the nervous system that arises from the brain stem and runs down either side of the neck. The vagus nerves accompany the major blood vessels of the neck (internal jugular veins) and innervate the heart and other viscera in the chest and abdominal cavities. They are partially responsible for the control of the heart rate and other vital functions. Sudden stimulation of the vagus nerves can cause immediate death due to cardiac failure, with the victim suddenly dropping dead. Some examples of sudden death by vagal stimulation can include a blow in the solar plexus, any form of pressure on the neck, and sudden dilatation of the neck of the womb (e.g. illegal abortion).

valency the combining power or bonding potential of an ATOM or GROUP, measured by the number of hydrogen ions (H^+, valency 1) that the atom could replace or combine with. In an IONIC compound, the charge on each ion represents the valency, e.g. in NaCl, both Na^+ and Cl^- have a valency of one. In COVALENT compounds, the valency is represented by the number of bonds that are formed, thus in carbon dioxide (CO_2), the carbon has a valency of 4 and oxygen 2 (*see* VALENCY ELECTRONS).

valency electrons the electrons present in the outermost shell of an atom of an element. Some elements always have the same number of valence electrons, e.g. hydrogen has one, ordinary oxygen has two, and calcium has two. The valence electrons of an atom are the ones involved in forming bonds with other atoms and are therefore shared, lost or gained when a compound or ION is formed.

valve structure that restricts flow through an opening, preventing back flow or ensuring unidirectional flow. There are valves in the heart (TRICUSPID), veins and lymphatic system. In botany, any part of a CAPSULE or dry fruit that releases its seeds when ripe.

Van der Waals' forces the weak, attractive force

between two neighbouring atoms. They are named after the Dutch physicist, Johannes van der Waals (1837–1923), who first discovered the phenomenon. In any atom, the electrons are continually moving and therefore have random distribution within the electron cloud of the atom. At any one moment, the electron cloud of an atom may be distorted so that a transient DIPOLE is produced. If two non-covalently bonded atoms are close enough together, the transient dipole in one atom will disturb the electron cloud of the other. This disturbance will create a transient dipole in the second atom, which will in turn attract the dipole in the first. It is the interaction between these transient dipoles that results in weak, non-specific Van der Waals' forces. Van der Waals' forces occur between all types of molecules, but they decrease in strength with increasing distance between the atoms or molecules.

variation the difference between individuals (plant or animal) within a population or species. If environmental conditions cause the variation, then it will not necessarily recur in succeeding generations. However, genetically caused variations will persist from one generation to the next. A species with wide genetic variations is more likely to survive changing conditions because the chances are increased that some members of the group will tolerate and adapt to any particular change.

varve a lacustrine deposit, near to ice sheets, of banded clays, silts and sands. The rhythmically banded sediments were deposited annually in lakes at the edge of ice sheets. Spring meltwaters bring new loads of sediment into the lake; the coarse particles are deposited quickly while the finer particles are only deposited from suspension later in the year. Since the glacial streams would refreeze in the winter, the sediment supply would cease until the following spring. This cyclic activity produces the banded effect, and each season accounts for one pale coarse band and one dark finer band. Varve deposits are very thin, and sixty or seventy years may be accommodated in one metre of sediment. Through counting varves, it is feasible to establish a timescale going back approximately 20,000 years.

vascular bundle a strand of conducting vascular tissue comprising XYLEM and PHLOEM separated in some plants by the cambium. It occurs in seed plants and the ferns and extends from the roots to the leaves.

vascular plants all plants with VASCULAR TISSUE— the pteridophytes, e.g. ferns, and spermatophytes, e.g. grasses, trees, shrubs, etc.

vascular system in plants, this is the VASCULAR TISSUE. In animals it is the network that circulates fluids through the body tissues. Most animals (except for protozoans and some simple invertebrates), have a vascular system for blood that participates in the movement of gases for respiration, nutrients, waste, etc. In vertebrates this system comprises the heart, veins, arteries and capillaries—a basic homeostatic system (*see* HOMEOSTASIS). The ECHINODERMATA have a water vascular system.

vascular tissue the XYLEM and PHLOEM that together form a tissue for transport of nutrients and water through the body of seed plants and ferns. The presence of vascular tissue enables such plants to grow to great heights and to dominate most land areas.

vas deferens tubes (a pair in mammals), that carry sperm from the TESTIS, through the urethra to the outside.

vas efferens small tubes or ducts that, in vertebrates, carry sperm from the seminiferous tubules to the outside (in the TESTIS) to the EPIDIDYMIS.

vasoconstriction the narrowing of blood vessels, especially ARTERIOLES, reducing the internal diameter and increasing blood pressure. The constriction is achieved through the contraction of SMOOTH MUSCLE in the arteriole walls under the control, in the main, of the SYMPATHETIC NERVOUS SYSTEM.

vasodilation the widening of blood vessels, especially ARTERIOLES, increasing the internal diameter and decreasing blood pressure. As with VASOCONSTRICTION, under the control, in the main, of the SYMPATHETIC NERVOUS SYSTEM.

vasomotor nerves nerves of the AUTONOMIC NERVOUS SYSTEM that control blood vessel diameter.

vasopressin (antidiuretic hormone, ADH) a peptide hormone, produced in the HYPOTHALAMUS, and secreted by the posterior PITUITARY GLAND. Vasopressin controls body fluids concentration by stimulating reabsorption of water in the KIDNEYS. ADH deficiency may lead to *Diabetes insipidus*, where urine production and excretion is abnormally high.

vector the term vector represents the plasmid used to carry a DNA segment into the host's cells or the organism that acts as a mechanism for transmitting a parasitic disease, e.g. mosquitoes are vectors of malaria.

vegetative propagation asexual reproduction in plants through the production of a separate structure, detached from the parent plant, e.g. TUBERS,

CORMS, BULBS, etc, and runners (as found in the strawberry plant). GRAFTS, buds and cuttings are artificial methods. Also refers to the reproduction of *Hydra* by budding (*see also* STOLON).

vein any thin-walled vessel that carries blood back from the body to the HEART. Veins contain few muscle fibres but have one-way VALVES that prevent backflow, thus enabling the blood to flow from body areas back to the heart.

veliger a planktonic molluscan larva. The foot, mantle and shell of the adult form are present.

vena cava one of the two major veins that empty into the right chamber of the HEART. The superior vena cava (SVC) carries the blood collected from the upper part of the body, e.g. neck and brain, while the inferior vena cava (IVC) carries the blood from the lower half of the body, e.g. liver, kidney and legs.

venation the pattern of veins in a leaf. Dicotyledons have a network while the leaves of monocotyledons have a parallel venation. The latter shows veins running parallel to the length of the leaf while the net (or reticular) pattern has a central vein with branches that may then branch further. Venation is also the term used to describe the veins in an insect's wing.

ventilation the rhythmic breathing, or external respiration, that is accomplished in different ways by different animals. Insects that are large and active create a movement of the TRACHEAE to pump air in and out of the SPIRACLES. In bony fish, the gills are ventilated by a current of water entering the mouth, passing through pharyngeal slits, over the gills and out via the back of the OPERCULUM. Amphibians use throat movements to maintain a fresh air flow, but birds have a complex system. In addition to lungs, birds have air sacs, mainly around the abdomen and neck. These are used in flight but also as pumps to assist air flow through the lungs. Further, the air flow through the lungs and air sacs is in one direction only. Mammals are the only animals to have a DIAPHRAGM to aid in ventilation. During inhaling/exhaling a volume of air called the *tidal volume* is taken in, about 500 ml in humans. The *vital capacity* is the maximum air volume that can be inhaled or exhaled (from 3,000 to 5,000 ml depending upon sex, stature, etc) but even then there is a *residual volume* of air remaining in the lungs.

ventral the surface of an animal or plant nearest to the ground. In animals such as man, it is the surface facing forward.

ventricle a major chamber of the HEART, which is thick-walled and muscular as it is the main pumping chamber. The outflow of the right ventricle is known as the PULMONARY ARTERY, which distributes blood to the lungs, and the outflow of the left ventricle is called the AORTA, which distributes blood to the head and the rest of the body.

venule a small blood vessel (between CAPILLARY and VEIN), usually with a valve to prevent back flow.

vermiform appendix *see* **appendix**.

vernalization the use of cold, when applied to seeds or seedlings, to ensure that they flower. Some cereals (and other plants) require a period of chilling in their early development to generate flowers, and are thus sown in the autumn. Artificial vernalization is used to enable winter cereals to be sown and then flower in one season.

vertebra (*plural* **vertebrae**) any of the bones in the vertebral column. In man there are groups of verte-brae: the cervical (7); thoracic (12); lumbar (5); sacral (5 fused) and 5 fused caudal vertebrae that form the coccyx. Each vertebra has an arch (neural arch) for passage of the SPINAL CORD and transverse processes, protrusions at the side.

vertebral column (spine, spinal column, backbone) the bony column made up of VERTEBRAE that runs near the DORSAL surface of vertebrates. It protects the SPINAL CORD and provides attachment for mus-cles. The vertebrae are separated by discs of CARTI-LAGE.

Vertebrata the major subphylum, within the phy-lum Chordata, that includes all animals with back-bones: fishes, amphibians, reptiles, birds and mam-mals. Vertebrates have well-formed sense organs, with the brain highly developed in many cases and enclosed by the skull. The circulatory system is closed with a pumping heart, and there is a complex excretory system involving the kidneys. Reproduc-tion is sexual and fertilization may be internal or external depending upon the species.

vesicle a small SAC in the CYTOPLASM of a cell that is usually filled with fluid. Vesicles occur in the GOLGI APPARATUS and are used to transport materials from the endoplasmic reticulum.

vestigial organ an organ that over evolutionary time has lost its use or importance and has reduced in size, for example some snakes show the remains of the pelvis and leg bones of their ancestors.

villus (*plural* **villi**) a tiny outgrowth from the sur-face of an organ or tissue that increases the availa-ble area. The small intestine contains great numbers, specialized for the absorption of food material. Each villus, up to a millimetre long in human beings, has blood and lymph vessels (*see also* MICRO-VILLUS).

virus the smallest microbe, which is completely parasitic as it is incapable of growing or reproducing outside the cells of its host. Most, but not all, viruses cause disease in plant, animal and even bacterial cells. Viruses are classified according to their nucleic acids and can contain double-stranded (DS) or single-stranded (SS) DNA or RNA. In infection, any virus must first bind to the host cells and then penetrate to release the viral DNA or RNA. The viral DNA or RNA then takes control of the cell's metabolic machinery to replicate itself, form new viruses, and then release the mature virus by either budding from the cell wall or rupturing and hence killing the cell. Some familiar examples of virus-induced diseases are herpes (double-stranded DNA), influenza (single-stranded RNA) and the retroviruses (single-stranded RNA, believed to cause AIDS and perhaps CANCER).

viscera referring to organs within the COELOMIC cavities. In human beings this includes the abdomen and thorax.

visual purple *see* **rhodopsin**.

vitamin any of a group of organic compounds that are required in small amounts in the diet, to maintain good health. Deficiencies lead to specific diseases. Vitamins form two groups: A, D, E and K are fat-soluble, while C (ASCORBIC ACID) and B (THIAMINE) are water-soluble.

vitamin A *or* **retinol** a fat soluble vitamin that must be included in the diet since vertebrates are unable to synthesize it. Vitamin A is a vital constituent of RHODOPSIN and deficiency can lead to night blindness and eventual total blindness. It is also involved in the maintenance of epithelial tissues. Vitamin A is available as a precursor in green vegetables and retinol is found in dairy products.

vitamin B (complex) a group of water-soluble vitamins that are coenzymes or components of coenzymes. Most animals have to acquire vitamin B through the diet, although many plants manufacture B vitamins. In addition to THIAMINE, RIBOFLAVIN, NICOTINIC ACID (niacin), PANTOTHENIC ACID, BIOTIN and FOLIC ACID, the complex contains B_6, pyridoxine that is involved in amino acid metabolism and is found in cereals, yoghurt, liver and milk. Vitamin B_{12} (cyanocobalamin) is a coenzyme in amino acid metabolism and is necessary for red blood cell production. It is found in meat (muscle), eggs and dairy products and anaemia results from a deficiency in this vitamin. Other components of the B complex include lipoic acid.

vitamin D a fat-soluble vitamin occurring as two steroid derivatives: D_2 or calciferol in yeast, and D_3 or cholecalciferol, the latter being produced by the action of sunlight on the skin (which contains a cholesterol derivative). The dietary source of vitamin D is fish-liver oils, eggs and dairy products and it is involved in blood calcium levels. The vitamin prompts increased calcium take-up in the gut, increasing the supply for the production of bone. Deficiency therefore causes bone deformities (rickets).

vitamin E *or* **tocopherol** a fat-soluble vitamin comprising several compounds that prevents damage to cell membranes. It is found in cereal grains and green vegetables.

vitamin K *or* **phylloquinone** a compound form; a fat-soluble vitamin that acts as a coenzyme in protein synthesis for blood clotting. Green vegetables and egg yolks contain vitamin K, but a deficiency (which can cause severe bleeding) is rare because a gut bacterium produces a form of the vitamin.

vitreous humour in the vertebrate eye, the jelly-like substance occurring in the space between the LENS and the RETINA.

viviparous a term describing any animal that gives birth to young that have developed inside its body. Viviparity is not restricted to mammals but also applies to some species of insect, e.g. the species of mite, *Acarophenox*, whose young develop by devouring and thus killing the mother.

vocal cords in air-breathing vertebrates, two membranes in the larynx that vibrate when expelled air is passed over them, thus producing sound. Muscles and cartilage control tension in the cords, which varies the sound created (*see also* LARYNX).

volt (V) the SI unit of potential difference. One volt is equal to one joule per coulomb of charge, i.e. $V = JC^{-1}$.

voltage the electrical energy that moves charge around a circuit. Voltage is the same as potential difference, and is thus measured in VOLTS. It is calculated between two given points on a circuit, and can be derived from the following equation: $V = d \times E$, where V = potential difference, d = difference between 2 points, E = electric field strength.

volume the space occupied by any object or substance. The volume of a liquid will depend on the amount of container space it occupies, but the volume of any gas will vary with pressure and temperature. Volume is measured in cm^3 or m^3. The volume of a cube, cuboid or cylinder is equal to the area of the base x height; the volume of a pyramid or cone is equal to 1/3 of the area of the base x height. The volume of a sphere is $4/3\pi r^3$.

volumetric analysis chemical analysis that uses standard solutions of known concentrations to calculate a particular constituent present in another solution, using titration.

voluntary muscle *see* **striated muscle**.

W

warfarin an anticoagulant used medically to thin the blood and in lethal doses as a rodenticide.

warmbloodedness *see* **homoiothermy**.

water a ubiquitous compound, hydrogen oxide (H_2O), which can occur as solid, liquid and gas phases. It forms a very large part of the Earth's surface and is vital to life. It occurs in all living organisms and has a remarkable combination of properties in its solvent capacity, chemical stability, thermal properties and abundance.

water potential the tendency of water to move by diffusion, osmosis or as vapour. At a pressure of one atmosphere, pure water is given a water potential value of zero, and hence all cells that water enters by osmosis have a water potential value less than zero. The water potential of any cell can be calculated using the following:

water potential = osmotic potential + pressure potential.

water table the level below which water saturates the available spaces in the ground. A spring or river is formed when, due to geological conditions, the water table rises above ground level.

Watson, James Dewey (1928–) an American molecular biologist who, along with his colleague, the English biologist Francis H. Crick (1916–), constructed a model revealing the structure of the DNA molecule. In 1962, they shared the Nobel prize for their work, and their double-helical model of DNA, showing a simple, repeating pattern of paired nucleic acid bases, suggested a means by which DNA replicates. Watson published an account of the discovery of DNA structure in his book, *The Double Helix* (1968).

watt (W) a unit of power that is the rate of work done at 1 JOULE per second, i.e. $1W = 1Js^{-1}$.

wax one of two types of solids or semi-solids. High molecular weight mixes of HYDROCARBONS are mineral waxes, e.g. paraffin wax. Plant or animal waxes are fatty acid ESTERS of alcohols and are protective, e.g. beeswax in the honeycomb, and the coating on leaves and fruits.

weight the gravitational force of attraction exerted by the earth on an object. As weight is a FORCE, its unit is the NEWTON (N). The weight of any object on earth can be calculated using:

W = mg m = mass (kg)

g = gravitational constant = $9.8ms^{-2}$

In everyday use, the term weight really refers to the mass of a person or object.

whalebone *or* **baleen** horny plates used on a feeding mechanism of certain whales (*see* CETACEA). The plates are suspended from the upper jaw on either side of the mouth and plankton are collected here when water is expelled through the mouth.

white blood cell *see* **leucocyte**.

white matter nerve tissue in the CENTRAL NERVOUS SYSTEM (brain and spinal cord) of vertebrates. It is composed primarily of nerve fibres in light coloured MYELIN SHEATHS.

wild type the term that originally signified the ALLELE possessed by most individuals in a natural population. It is now also used in the laboratory as being the stock from which mutants are produced.

wood the secondary XYLEM and fibres of many perennial plants that form the hard structural (heartwood) and water-conducting (sapwood) tissue. *Hardwood* refers to the ANGIOSPERMAE, e.g. mahogany, and pine is a typical softwood (*see* GYMNOSPERMAE).

X

xanthophyll a carotenoid pigment that produces the typical brown and yellow colours of autumn leaves.

X-chromosome one kind of CHROMOSOME that is involved in the sex determination of an individual. A woman has a pair of X-chromosomes, whereas a man has one X-chromosome and one Y-chromosome. There are many GENES on the X-chromosome that have nothing to do with the sex of the individual. For example, red-green colour blindness is determined by a RECESSIVE gene on the X-chromosome. If a woman carrying this gene has a son (X,Y) then he will inherit colour blindness as the Y-chromosome will have no corresponding gene to suppress the effect. If she has a daughter (X,X), and the father has normal vision, then the recessive gene is still inherited but its effect is suppressed as the X-chromosome from the father will carry the dominant gene for normal vision.

xerophyte a plant that is able to live in an arid area with little water in the soil. Such plants have special structures such as swellings to conserve water and thick CUTICLE or leaves to reduce transpirational loss. A typical example is the desert cactus.

X-rays the part of the electromagnetic spectrum

with a wavelength range of approximately 10^{-12} to 10^{-9}m and a frequency range of 10^{17} to 10^{21}Hz. X-rays are produced when electrons moving at high speed are absorbed by a target. The resultant waves will penetrate solids to varying degrees, dependent on the density of the solid. Hence X-rays of certain wavelengths will penetrate flesh, but not bone or other more dense materials. X-rays serve both therapeutic and diagnostic functions in medicine and are deployed in many areas of industry where inspection of hidden, inaccessible objects is necessary.

xylem the plant tissue that transports water and nutrients, and because the cell walls are lignified, xylem also functions in support. Flowering plants have hollow tubes (*vessels*) joined end to end. Primary xylem is formed by differentiation from the PROCAMBIUM and secondary xylem is formed by the cambium (a tissue of actively dividing cells responsible for increasing the diameter of a plant).Wood is made up mainly of secondary xylem.

Y

Y-chromosome the small chromosome that carries a dominant gene for maleness. All normal males have 22 matched pairs of chromosomes and one unmatched pair, one large X-chromosome and one small Y-chromosome. The X-chromosome, which is inherited from the mother, carries many more genes than the Y-chromosome, which is inherited from the father. During sexual reproduction, the mother must contribute one X-chromosome, but the father can contribute either an X or Y-chromosome. The effect of the Y-chromosome is that a testis develops in the embryo instead of an ovary. Thus the sex of the resulting offspring is dependent on the father's contribution—female (X,X) or male (X,Y).

yeast unicellular micro-organisms that form a fungus. Yeast cells can be circular or oval in shape and reproduce by spore formation. The enzymes secreted by yeasts are used in brewing and baking industries as they can convert sugars into alcohol and carbon dioxide.

yolk an embryonic food store in an egg. It consists mainly of protein and/or fats.

yolk sac an extra-embryonic membrane in vertebrates (mainly reptiles and birds) that contains yolk and is attached to the ventral surface of the embryo.

As the yolk is absorbed, the sac merges with the embryo. In mammals there is fluid not yolk in the sac, but it is homologous with the yolk sac of birds, etc, and is the site of the early formation of blood cells.

Z

zoology a branch of biology that involves the study of animals. Subjects studied include anatomy, physiology, embryology, evolution, and the geographical distribution of animals.

zooplankton that part of the PLANKTON composed of animals, which are the larvae, eggs or adults of many phyla. The numbers are enormous in surface waters.

zwitterion the predominant form of an AMINO ACID when surrounded by a neutral solution (pH 7). The structure of a zwitterion and amino acid differ in that the zwitterion exists as a dipolar ion—the carboxyl (-COOH) group of the amino acid loses a hydrogen atom to form -COO$^-$, and the amino group (-NH$_2$) gains a hydrogen atom to form -NH$_3$$^+$.

zygote the cell immediately produced by the fusion of male and female germ cells (GAMETES) during the initial stage of FERTILIZATION. The zygote is a DIPLOID cell, formed by the fusion of the haploid male gamete and the haploid female gamete.

zymogen an inactive form of an ENZYME. Most zymogens are inactive precursors of pancreatic enzymes, which are involved in protein digestion. Synthesis of these digestive enzymes as zymogens prevents the unwanted digestion of the tissue in which the enzyme was made. The zymogen becomes activated by chemical modifications to form its functional form when it reaches its site of function, e.g. the enzyme chymotrypsin (digests protein) is synthesized in the pancreas as the zymogen, chymotrypsinogen, and becomes activated only when it reaches its destination, the small intestine.

zymurgy a branch of chemistry that involves the study of FERMENTATION processes.

Periodic Table

Group								TRANSITION ELEMENTS									
1A	2A	3B	4B	5B	6B	7B	8	8	8	1B	2B	3A	4A	5A	6A	7A	0
H 1																	He 2
Li 3	Be 4											B 5	C 6	N 7	O 8	F 9	Ne 10
Na 11	Mg 12											Al 13	Si 14	P 15	S 16	Cl 17	Ar 18
K 19	Ca 20	Sc 21	Ti 22	V 23	Cr 24	Mn 25	Fe 26	Co 27	Ni 28	Cu 29	Zn 30	Ga 31	Ge 32	As 33	Se 34	Br 35	Kr 36
Rb 37	Sr 38	Y 39	Zr 40	Nb 41	Mo 42	Tc 43	Ru 44	Rh 45	Pd 46	Ag 47	Cd 48	In 49	Sn 50	Sb 51	Te 52	I 53	Xe 54
Cs 55	Ba 56	La¹ 57	Hf 72	Ta 73	W 74	Re 75	Os 76	Ir 77	Pt 78	Au 79	Hg 80	Tl 81	Pb 82	Bi 83	Po 84	At 85	Rn 86
87	88	89															

¹ Lanthanides	La 57	Ce 58	Pr 59	Nd 60	Pm 61	Sm 62	Eu 63	Gd 64	Tb 65	Dy 66	Ho 67	Er 68	Tm 69	Yb 70	Lu 71
² Actinides	Ac 89	Th 90	Pa 91	U 92	Np 93	Pu 94	Am 95	Cm 96	Bk 97	Cf 98	Es 99	Fm 100	Md 101	No 102	Lr 103

APPENDIX 2

Element Table

Element	Symbol	Atomic Number	Relative Atomic Mass*
Actinium	Ac	89	{227}
Aluminium	Al	13	26.9815
Americium	Am	95	{243}
Antimony	Sb	51	121.75
Argon	Ar	18	39.948
Arsenic	As	33	74.9216
Astatine	At	85	{210}
Barium	Ba	56	137.34
Berkelium	Bk	97	{247}
Beryllium	Be	4	9.0122
Bismuth	Bi	83	208.98
Boron	B	5	10.81
Bromine	Br	35	79.904
Cadmium	Cd	48	112.40
Caesium	Cs	55	132.905
Calcium	Ca	20	40.08
Californium	Cf	98	{251}
Carbon	C	6	12.011
Cerium	Ce	58	140.12
Chlorine	Cl	17	35.453
Chromium	Cr	24	51.996
Cobalt	Co	27	58.9332
Copper	Cu	29	63.546
Curium	Cm	96	{247}
Dysprosium	Dy	66	162.50

Element	Symbol	Atomic Number	Relative Atomic Mass*
Einsteinium	Es	99	{254}
Erbium	Er	68	167.26
Europium	Eu	63	151.96
Fermium	Fm	100	{257}
Fluorine	F	9	18.9984
Francium	Fr	87	{223}
Gadolinium	Gd	64	157.25
Gallium	Ga	31	69.72
Germanium	Ge	32	72.59
Gold	Au	79	196.967
Hafnium	Hf	72	178.49
Helium	He	2	4.0026
Holmium	Ho	67	164.930
Hydrogen	H	1	1.00797
Indium	In	49	1114.82
Iodine	I	53	126.9044
Iridium	Ir	77	192.2
Iron	Fe	26	55.847
Krypton	Kr	36	83.80
Lanthanum	La	57	138.91
Lawrencium	Lr	103	{257}
Lead	Pb	82	207.19
Lithium	Li	3	6.939
Lutetium	Lu	71	174.97
Magnesium	Mg	12	24.305
Manganese	Mn	25	54.938
Mendelevium	Md	101	{258}

Element	Symbol	Atomic Number	Relative Atomic Mass*
Mercury	Hg	80	200.59
Molybdenum	Mo	42	95.94
Neodymium	Nd	60	144.24
Neon	Ne	10	20.179
Neptunium	Np	93	{237}
Nickel	Ni	28	58.71
Niobium	Nb	41	92.906
Nitrogen	N	7	14.0067
Nobelium	No	102	{255}
Osmium	Os	76	190.2
Oxygen	O	8	15.9994
Palladium	Pd	46	106.4
Phosphorus	P	15	30.9738
Platinum	Pt	78	195.09
Plutonium	Pu	94	{244}
Polonium	Po	84	{209}
Potassium	K	19	39.102
Praseodymium	Pr	59	140.907
Promethium	Pm	61	{145}
Protactinium	Pa	91	{231}
Radium	Ra	88	{226}
Radon	Rn	86	{222}
Rhenium	Re	75	186.20
Rhodium	Rh	45	102.905
Rubidium	Rb	37	85.47
Ruthenium	Ru	44	101.07
Samarium	Sm	62	150.35

Element	Symbol	Atomic Number	Relative Atomic Mass*
Scandium	Sc	21	44.956
Selenium	Se	34	78.96
Silicon	Si	14	28.086
Silver	Ag	47	107.868
Sodium	Na	11	22.9898
Strontium	Sr	38	87.62
Sulphur	S	16	32.064
Tantalum	Ta	73	180.948
Technetium	Tc	43	{97}
Tellurium	Te	52	127.60
Terbium	Tb	65	158.924
Thallium	Tl	81	204.37
Thorium	Th	90	232.038
Thulium	Tm	69	168.934
Tin	Sn	50	118.69
Titanium	Ti	22	47.90
Tungsten	W	74	183.85
Uranium	U	92	238.03
Vanadium	V	23	50.942
Xenon	Xe	54	131.30
Ytterbium	Yb	70	173.04
Yttrium	Y	39	88.905
Zinc	Zn	30	65.37
Zirconium	Zr	40	91.22

*Values of the *Relative Atomic Mass* in brackets refer to the most stable, known, isotope.

Elements listed by symbol

Symbol	Element	Symbol	Element
Ac	Actinium	Mn	Manganese
Ag	Silver	Mo	Molybdenum
Al	Aluminium	N	Nitrogen
Am	Americium	Na	Sodium
Ar	Argon	Nb	Niobium
As	Arsenic	Nd	Neodymium
At	Astatine	Ne	Neon
Au	Gold	Ni	Nickel
B	Boron	No	Nobelium
Ba	Barium	Np	Neptunium
Be	Beryllium	O	Oxygen
Bi	Bismuth	Os	Osmium
Bk	Berkelium	P	Phosphorus
Br	Bromine	Pa	Protactinium
C	Carbon	Pb	Lead
Ca	Calcium	Pd	Palladium
Cd	Cadmium	Pm	Promethium
Ce	Cerium	Po	Polonium
Cf	Californium	Pr	Praseodymium
Cl	Chlorine	Pt	Platinum
Cm	Curium	Pu	Plutonium
Co	Cobalt	Ra	Radium
Cr	Chromium	Rb	Rubidium
Cs	Caesium	Re	Rhenium
Cu	Copper	Rh	Rhodium
Dy	Dysprosium	Rn	Radon
Er	Erbium	Ru	Ruthenium
Es	Einsteinium	S	Sulphur
Eu	Europium	Sb	Antimony
F	Fluorine	Sc	Scandium
Fe	Iron	Se	Selenium
Fm	Fermium	Si	Silicon
Fr	Francium	Sm	Samarium
Ga	Gallium	Sn	Tin
Gd	Gadolinium	Sr	Strontium
Ge	Germanium	Ta	Tantalum
H	Hydrogen	Tb	Terbium
He	Helium	Tc	Technetium
Hf	Hafnium	Te	Tellurium
Hg	Mercury	Th	Thorium
Ho	Holmium	Ti	Titanium
I	Iodine	Tl	Thallium
In	Indium	Tm	Thulium
Ir	Iridium	U	Uranium
K	Potassium	V	Vanadium
Kr	Krypton	W	Tungsten
La	Lanthanum	Xe	Xenon
Li	Lithium	Y	Yttrium
Lr	Lawrencium	Yb	Ytterbium
Lu	Lutetium	Zn	Zinc
Md	Mendelevium	Zr	Zirconium
Mg	Magnesium		

APPENDIX 3

THE GREEK ALPHABET

Name	Capital	Lower Case	English Sound
alpha	A	α	a
beta	B	β	b
gamma	Γ	γ	g
delta	Δ	δ	d
epsilon	E	ε	e
zeta	Z	ζ	z
eta	H	η	e
theta	Θ	θ	th
iota	I	ι	i
kappa	K	κ	k
lambda	Λ	λ	l
mu	M	μ	m
nu	N	ν	n
xi	Ξ	ξ	x
omicron	O	ο	o
pi	Π	π	p
rho	P	ρ	r
sigma	Σ	σ	s
tau	T	τ	t
upsilon	Y	υ	u
phi	Φ	φ	ph
chi	X	χ	kh
psi	Ψ	ψ	ps
omega	Ω	ω	o

APPENDIX 4

The International System of Units (SI units)

Quantity	Symbol	Unit	Symbols
acceleration	a	metres per second squared	ms^{-2} or m/s^2
area	A	square metre	m^2
capacitance	C	farad	F ($1F = 1AsV^{-1}$)
charge	Q	coulomb	C ($1C = 1As$)
current	I	ampere	A
density	ρ	kilograms per cubic metre	kgm^{-3} or kg/m^3
force	F	newton	N ($1N = 1\ kg\ ms^{-2}$)
frequency	f	hertz	Hz ($1Hz = 1s^{-1}$)
length	l	metre	m
mass	m	kilogram	kg
potential difference	V	volt	V ($1V = 1JC^{-1}$ or WA^{-1})
power	P	watt	W ($1W = 1Js^{-1}$)
resistance	R	ohm	Ω ($1\Omega = 1VA^{-1}$)
specific heat capacity	c	joules per kilogram kelvin	$Jkg^{-1}\ K^{-1}$
temperature	T	kelvin	L
time	t	second	s
volume	V	cubic metre	m^3
velocity	v	metres per second	ms^{-1} or m/s
wavelength	λ	metre	m
work, energy	W, E	joule	J ($1J = 1Nm$)

APPENDIX 4 (cont.)

Useful prefixes adopted with SI units

Prefix	Symbol	Factor
tera	T	10^{12}
giga	G	10^{9}
mega	M	10^{6}
kilo	k	10^{3}
hecto	h	10^{2}
deda	da	10^{1}
deci	d	10^{-1}
centi	c	10^{-2}
milli	m	10^{-3}
micro	μ	10^{-6}
nano	n	10^{-9}
pico	p	10^{-12}
femto	f	10^{-15}
atto	a	10^{-18}

APPENDIX 5

Geological Time Scale

Eon	Era	Sub-era	Period	Epoch	Millions of years since the start
PHANEROZOIC	Cenozoic	Quaternary	Pleistogene	Holocene	0.01
				Pleistocene	2.0
		Tertiary	Neogene	Pliocene	5.1
				Miocene	24.6
			Palaeogene	Oligocene	38
				Eocene	55
				Palaeocene	65
	Mesozoic		Cretaceous		144
			Jurassic		213
			Triassic		248

Geological Time Scale

Eon	Era	Sub-era	Period	Epoch	Millions of years since the start
PHANEROZOIC	Palaeozoic	Upper Palaeozoic	Permian		286
			Carboniferous		360
			Devonian		408
		Lower Palaeozoic	Silurian		438
			Ordovician		505
			Cambrian		590
PROTEROZOIC P R E					2500
ARCHAEAN C A M B					4000
PRISCOAN R I A N					4600

The Solar System

Planet	Diameter at the Equator km	Mass relative to the Earth[1]	Average distance from Sun km[6]	The planet's "year"
Mercury	44840	0.054	57.91	87.969 days
Venus	12300	0.8150	108.21	224.701 days
Earth	12756	1.000	149.60	365.256 days
Mars	6790	0.107	227.94	686.980 days
Jupiter	142700	317.89	778.34	11.86 years
Saturn	120800	95.14	1427.01	29.46 years
Uranus	50800	14.52	2869.6	84.0 years
Neptune	48600	17.46	4496.7	164.8 years
Pluto	3500	0.1 (approx)	5907	248.4 years
Sun	1392000	332 958		
Moon	3476	0.0123		

[1] The mass of the Earth is 5.976×10^{24} kg

448